# MACHINE RECONSTRUCTION OF HUMAN CONTROL STRATEGIES

Frontiers in Artificial Intelligence and Applications

Volume 99

*Published in the subseries*

Dissertations in Artificial Intelligence

*Under the Editorship of the ECCAI Dissertation Board*
*Proposing Board Member: Franc Solina*

*Recently published in this series:*

Vol. 98. H. Fujita and P. Johannesson (Eds.), New Trends in Software Methodologies, Tools and Techniques – Proceedings of Lyee_W03: The Second International Workshop on Lyee Methodology
Vol. 97. H.U. Hoppe et al. (Eds.), Artificial Intelligence in Education – Shaping the Future of Learning through Intelligent Technologies
Vol. 96. S. Handschuh and S. Staab (Eds.), Annotation for the Semantic Web
Vol. 95. B. Omelayenko and M. Klein (Eds.), Knowledge Transformation for the Semantic Web
Vol. 94. H. Jaakkola et al. (Eds.), Information Modelling and Knowledge Bases XIV
Vol. 93. K. Wang, Intelligent Condition Monitoring and Diagnosis Systems – A Computational Intelligence Approach
Vol. 92. V. Kashyap and L. Shklar (Eds.), Real World Semantic Web Applications
Vol. 91. F. Azevedo, Constraint Solving over Multi-valued Logics – Application to Digital Circuits
Vol. 90. In preparation
Vol. 89. T. Bench-Capon et al. (Eds.), Legal Knowledge and Information Systems – JURIX 2002: The Fifteenth Annual Conference
Vol. 88. In preparation
Vol. 87. A. Abraham et al. (Eds.), Soft Computing Systems – Design, Management and Applications
Vol. 86. R.S.T. Lee and J.H.K. Liu, Invariant Object Recognition based on Elastic Graph Matching – Theory and Applications
Vol. 85. J.M. Abe and J.I. da Silva Filho (Eds.), Advances in Logic, Artificial Intelligence and Robotics – LAPTEC 2002
Vol. 84. H. Fujita and P. Johannesson (Eds.), New Trends in Software Methodologies, Tools and Techniques – Proceedings of Lyee_W02
Vol. 83. V. Loia (Ed.), Soft Computing Agents – A New Perspective for Dynamic Information Systems
Vol. 82. E. Damiani et al. (Eds.), Knowledge-Based Intelligent Information Engineering Systems and Allied Technologies – KES 2002
Vol. 81. J.A. Leite, Evolving Knowledge Bases – Specification and Semantics
Vol. 80. T. Welzer et al. (Eds.), Knowledge-based Software Engineering – Proceedings of the Fifth Joint Conference on Knowledge-based Software Engineering
Vol. 79. H. Motoda (Ed.), Active Mining – New Directions of Data Mining
Vol. 78. T. Vidal and P. Liberatore (Eds.), STAIRS 2002 – STarting Artificial Intelligence Researchers Symposium
Vol. 77. F. van Harmelen (Ed.), ECAI 2002 – 15th European Conference on Artificial Intelligence
Vol. 76. P. Sinčák et al. (Eds.), Intelligent Technologies – Theory and Applications
Vol. 75. I.F. Cruz et al. (Eds.), The Emerging Semantic Web – Selected Papers from the first Semantic Web Working Symposium

ISSN 0922-6389

# Machine Reconstruction of Human Control Strategies

Dorian Šuc

*Faculty of Computer and Information Science, University of Ljubljana, Ljubljana, Slovenia*

IOS Press

Ohmsha

Amsterdam • Berlin • Oxford • Tokyo • Washington, DC

ISBN 1 58603 376 X (IOS Press)
ISBN 4 274 90617 5 C3055 (Ohmsha)
Library of Congress Control Number: 2003110988

*Publisher*
IOS Press
Nieuwe Hemweg 6B
1013 BG Amsterdam
The Netherlands
fax: +31 20 620 3419
e-mail: order@iospress.nl

*Distributor in the UK and Ireland*
IOS Press/Lavis Marketing
73 Lime Walk
Headington
Oxford OX3 7AD
England
fax: +44 1865 75 0079

*Distributor in the USA and Canada*
IOS Press, Inc.
5795-G Burke Centre Parkway
Burke, VA 22015
USA
fax: +1 703 323 3668
e-mail: iosbooks@iospress.com

*Distributor in Japan*
Ohmsha, Ltd.
3-1 Kanda Nishiki-cho
Chiyoda-ku, Tokyo 101-8460
Japan
fax: +81 3 3233 2426



PRINTED IN THE NETHERLANDS

# Preface

This book is about an investigation into the fascinating task of reconstructing sub cognitive human skill from observations of the human performance. What makes this problem particularly interesting is the fact that humans execute their skill without "really knowing" it. Crane operators, air pilots and monocycle acrobats may *perform* flawlessly, but they are only able to *describe* their skills vaguely. It has been shown that such descriptions of human skill typically lack completeness and correctness to be translated into a computer program that would reliably carry out the control task.

Given that skill is hard to access through introspection, one idea is to apply techniques of machine learning to extract operational descriptions of skill from operator's performance data. This idea occurred already in the 1960's, probably for the first time in the work of P. Donaldson, and independent of this in the well-known work by D. Michie and R. Chambers. A revived wave of interest in this topic appeared in the 1990's when new techniques of machine learning became available, with contributions by D. Michie, C. Sammut and T. Urbančič, among others. This area also became known as "behavioural cloning", because the goal is to produce a clone of the operator's skill that exhibits a similar control behaviour. In fact, the task of behavioural cloning is more ambitious than mere reproduction of behaviour: the aim also is to obtain an explicit insight into the skill, to understand how the operator essentially does it.

However, it turned out that straightforward application of machine learning tools to operator performance data had difficulties. Among other difficulties, the yield of cloning, that is, the proportion of successful clones among all clones induced from data, was low. In one comprehensive study in the crane domain, the yield was reported to be well below 10%. This is where the innovations by Dorian Šuc, described in the present book, give most spectacular improvements. Šuc's approach immediately increased the yield in the same domain to over 90%. This improvement becomes even more striking when compared with classification accuracy improvements reported of new methods in machine learning, which are typically in the order of no more than a few percent, say from 65 to 67%.

Šuc's approach comprises two main ideas:

1. The decomposition of the learning of control into two learning tasks: first, the learning of operator's trajectory, and second, the learning of the system's dynamics. A combination of the two learned models results in an "indirect controller". Such controllers are called *indirect* because system's states are mapped into controller's actions indirectly through generalised control trajectories.

2. The learning of "qualitative operator's trajectories". Šuc's system, QUIN induces *qualitative trees*, that is, tree-structured sets of constraints on operator's control trajectory. Such qualitative trees proved useful both in the optimisation and the understanding of operator's control strategies.

I expect this book will be of interest to researchers in Artificial Intelligence, particularly in Machine Learning and Data Mining, to researchers concerned with the psychological aspects of human skill, and to practitioners that seek unconventional tools for the development of controllers, in particular for capturing and mechanising human skills. Šuc's program QUIN has recently been applied within the European research project Clockwork to the task of reverse engineering of industrial controllers.

This book is based on Dorian Šuc's Ph.D. thesis. Supervising this thesis gave me a lot of pleasure, following the development of many interesting ideas and surprising results. I believe this work well deserved the 2001 ECCAI award for the best Ph.D. thesis in AI in Europe.

Ivan Bratko
Ljubljana, July 2003

# Abstract of the Thesis

Complex dynamic systems are usually controlled by the operators who acquire their skill through years of experience. Typically, such a control skill is sub-cognitive and hard to reconstruct through introspection. The operators cannot completely describe their own skill, however they can demonstrate it. Therefore, an attractive approach to human control skill reconstruction involves machine learning from the operator's execution traces. The goal is to induce a model of the operator's skill, a control strategy that helps to understand the skill and can be used to control the system.

Behavioural cloning is an approach to such skill reconstruction. In the "original" approach to behavioural cloning a strategy is induced as a direct mapping from system's states to actions in a form of a decision or regression tree. This thesis develops new ideas to tackle problems that have been generally observed with this approach to human skill reconstruction.

One idea is to decompose the learning problem and induce goal-directed strategies that consider learned models of system's dynamics. A generalized operator's trajectory that can be seen as a continuously changing subgoal is introduced. This improves the robustness of the resulting controllers.

Another idea, which is also relevant to the comprehensibility, is to induce *qualitative* models of human control skill. It is shown that such qualitative strategies provide an insight into the operator's control skill. On the basis of our experiments, we argue that qualitative strategies can capture important and non-trivial aspects of human control skill. Qualitative strategies also open other new perspectives of the reconstruction of human control skill, such as the reconstruction of individual differences in operators' control styles.

These ideas were implemented and evaluated in dynamic domains including a container crane, a double pendulum referred to as the acrobot, and bicycle riding. To induce a qualitative control strategies program, QUIN (Qualitative Induction) for learning qualitative constraint trees from numerical examples was developed.

# Acknowledgements

This book is based on my Ph.D. thesis. It was supervised by Ivan Bratko and has greatly benefited from his curiosity, sparkling ideas, and the ability to see the problems and solutions from a different perspective. Working with him was a great pleasure. I'm grateful for his guidance and constant encouragement over the years. I'm also grateful to Tanja Urbančič for useful discussions, and Claude Sammut for comments that improved the quality of my thesis.

Marko Grobelnik was responsible for raising my interest in computer science and Artificial Intelligence. He was always willing to share and enthusiastically discuss the ideas. I would also like to thank Tatjana Zrimec, Dunja Mladenič, Igor Kononenko, Blaž Zupan and other colleagues from the Artificial Intelligence Lab in Ljubljana for their comments and for making my work enjoyable.

Dorian Šuc
July 2003

# Contents

# Chapter 1

# Introduction

## 1.1 Motivation

Controlling a complex dynamic system, such as a plane or a crane, usually requires a skilled operator. Typically, such a control skill is acquired through years of experience and is sub-cognitive. Thus, it is hard to access and difficult to rationally reconstruct through introspection. One motivation for such skill reconstruction is in its use in the development of an automatic controller that could assist the operator, or control the system where working conditions are poor or tiring. Preferably, to control the system even in some way better than the operator. Another motivation for skill reconstruction is to transfer the experience of a skilled operator to other operators and to reduce the training time of novice operators. Understanding the control skill might help operators to perform the task more reliably, effectively and safely in many different and sometimes unexpected circumstances.

A machine learning approach to such skill reconstruction is behavioural cloning. The goal is not only to induce a successful controller, but also to achieve a better understanding of the operator's sub-cognitive skill. For this reason a symbolic model of the operator's skill is induced from execution traces of his/her control performance. Several studies in recent years show that qualitative models are often closer to human thinking and better suited for some tasks than the traditional quantitative, or numerical models. Therefore, qualitative modelling of human control skill is also a tempting idea.

## 1.2 Skill Reconstruction and Behavioural Cloning

The motivation for skill reconstruction is understanding the underlying operator's skill and its use in the development of an automatic controller. This is particularly interesting in the case of dynamic systems that are too complex for classical control theory solutions. Controllers for such systems are often designed using approaches like fuzzy control (Lee, 1990) and different kinds of machine learning. Approaches like neural nets (Hunt et al., 1992), reinforcement learning (Sutton and Barto, 1998) and genetic algorithms (Renders and Nordvik, 1992) do typically not use prior knowledge about the system to be controlled. This results in time-consuming experimentation with the dynamic system, a low success rate of learning and the lack of interpretability of learned controllers. Humans, however, rarely attempt to learn from scratch. They extract initial biases, as well as strategies, from their prior knowledge of the system or from a demonstration of experienced operators.

The idea of *behavioural cloning* (a term introduced by Donald Michie (1993) is to make use of the operator's skill in the development of an automatic controller. A skilled operator's execution traces are used as the examples for machine learning to reconstruct the underlying control strategy that the operator executes subconsciously. The goal of behavioural cloning is not only to induce a successful controller, but also to achieve a better understanding of the human operator's subconscious skill. Such rule-based acquisition of the real-time control skill was first described by Michie and Chambers (1969). An early work of learning by imitation to control an unstable system is also the work of Donaldson (1964), however, these systems did not provide a reusable model of control skill. Behavioural cloning was later successfully used in problem domains as pole balancing (Michie et al., 1990), piloting (Sammut et al., 1992), production line scheduling (Kerr and Kibira, 1994) and container cranes (Urbančič, 1994). Detailed reviews of these and more recent experiments in behavioural cloning are also in (Bratko et al., 1998; Sammut, 1996; Camacho, 2000).

Controllers, also called clones, were usually induced as a direct mapping from states to control actions in the form of decision or regression trees. Such controllers, induced with this approach to behavioural cloning (here referred to as the *original approach*), do provide some insight into the control strategy, but as stated by Urbančič and Bratko (1994), and Michie (1995), they in general lack conceptual structure that would clearly capture the causal relations in the domain and the goal structure of the control strategy. In the experiments in (Šuc and Bratko, 1997b; Šuc and Bratko, 1999b; Šuc and Bratko, 1999a) taking into account causality and dynamics resulted in improved robustness of induced controllers.

Because of their ability to flexibly accommodate background knowledge, inductive logic programming techniques (Muggleton, 1992) would seem to be more appropriate. As suggested in an early work with behavioural cloning (Sammut et al., 1992) this might also improve transparency of the induced rules. Some experiments with inductive logic programming have been performed in the piloting domain (Camacho, 1995; Camacho, 2000) but so far the results have generally not been better than those with decision or regression trees. However, there are indications (Camacho, 2000) that expert-provided constraints on the hypothesis language can improve the transparency of the clones.

Although successful clones have been induced using standard machine learning techniques in different domains, Urbančič and Bratko (1994) also report on difficulties with the original approach to behavioural cloning. These problems are summarized by Michie (1995) as follows:

- "The present approach lacks robustness in that they do not guarantee inducing with high probability a successful clone from given data.
- Typically, the induced clones are not sufficiently robust with respect to changes in the control task.
- Although the clones do provide some insight into the control strategy, they in general lack conceptual structure that would clearly capture the causal relations in the domain and the goal structure of the control strategy."

Further experience along similar lines was reported by several authors including the following. The observation that the clone induction process typically has low yield, i.e. that the proportion of successful controllers among all the induced clones is typically well below 50%, was reported in experiments with piloting (Camacho, 1994), double pendulum acrobot

(Miletić, 1997; Vladušič, 2000) and container cranes (Siegl, 1995; Juvan, 2000). An approach that addresses this problem is proposed in (Camacho, 1998; Camacho, 2000) and uses a wrapper (John et al., 1994) as a method to tune the parameters of the machine learning algorithm. Still, producing a successful controller requires testing of different clones (induced with different parameters) on the simulator. When referring to the *yield* of cloning in this thesis, we mean the probability of inducing a successful clone *without* having access to the controlled system or its simulator (this excludes wrapper-based approaches).

The observation that the clones are not robust with respect to small changes in the control task was also reported in experiments with piloting (Camacho, 1995; Camacho, 2000) and container cranes (Urbančič, 1994; Urbančič and Bratko, 1994; Bratko and Urbančič, 1999). This phenomenon can be illustrated with the observation from the piloting domain (Camacho, 1994) where decision trees were used to induce auto-pilots. The clones that successfully accomplished the original task were then tested with different start states of the plane. The author reports: "On the more complex manoeuvres like turn and align the accuracy reaches a minimum value of 0.0%."

The difficulty of the clones' comprehensibility was observed in many studies. As reported in (Bain and Sammut, 1999; Urbančič, 1994; Camacho, 2000) the clones induced with the original approach to behavioural cloning are often large and difficult to understand. The induced trees often have several hundreds of leaves and are, as such, not acceptable models of human skill (Bratko and Urbančič, 1999). As noted in (Bratko et al., 1995; Šuc and Bratko, 1997b; Šuc and Bratko, 1997a; Bratko et al., 1998), the clones are purely reactive and inadequately structured as conceptualizations of the human skill and lack typical elements of human control strategies such as goals, subgoals, phases and causality.

Donald Michie (1995) noted that "learning from exemplary performance requires more than mindless imitation" and suggested that a kind of a problem decomposition could be helpful for behavioural cloning. Several studies (Michie and Sammut, 1995; Stirling, 1995; Camacho, 1998; Bain and Sammut, 1999; Bratko and Urbančič, 1999; Camacho, 2000) followed this idea and addressed the stated problems of the original approach to behavioural cloning. These approaches are briefly discussed in Section 3.5. The solutions proposed in this thesis are different in many aspects.

To improve the robustness of induced controllers and the yield of cloning a decomposition of the cloning problem into two learning problems is proposed: the learning of the operator's control trajectories and the learning of the system's dynamics. In this way, goal-directed clones that consider the system's dynamics and causality in the system are induced. The generalized operator's trajectory can be seen as a continuously changing subgoal and, as such, provides an insight into the operator's skill.

Another idea that deals with the comprehensibility of the induced strategies is to induce qualitative models of human control skill. Several studies (Bratko et al., 1989; Falkenhainer and Forbus, 1990; Williams, 1990) in recent years show that qualitative models are better suited than the traditional quantitative or numerical models for several tasks. Qualitative models were also applied to reasoning about control. For example, by applying qualitative simulation to the analysis of heterogeneous fuzzy controllers (Kuipers and Åström, 1994), deriving a qualitative control rule from a qualitative differential equations model (Bratko, 1995) or by the use of a qualitative explanation of the control skill (DeJong, 1994). To our knowledge, no study has yet addressed the problem of inducing qualitative models of human control skill.

## 1.3 Organization of the Thesis

Chapter 2 describes the learning problem for skill reconstruction, dynamic domains used in experiments, and some standard machine learning algorithms that are used later in the thesis. In Chapter 3 the learning problem decomposition for behavioural cloning is proposed. Then experiments are presented and possible advantages of the proposed decomposition are discussed. Chapter 4 gives experiments with skill reconstruction in dynamic domains of container cranes and double pendulum referred to as acrobot. These experiments are a motivation to learn qualitative models of human control skill. In Chapter 5 algorithm QUIN for learning of qualitative constraint trees from numerical examples is developed. QUIN is applied to learning of qualitative control strategies in Chapter 6. Chapter 7 discusses the role of domain knowledge and user intervention. Finally, the thesis concludes with an overview of the presented work and the results.

## 1.4 Contributions of the Thesis

The contributions of the thesis are:

- A decomposition of the behavioural cloning problem into two learning problems: the learning of constraints on the operator's control trajectories and the learning of the system's dynamics. In the experiments carried out in the thesis, the decomposition improves the yield of cloning and the robustness of induced clones. A plausible explanation of why this decomposition works better than the original approach, where a controller is induced as a direct mapping from states to actions is given. The framework facilitates the domain knowledge.

- Modelling the operator's control skill with symbolic, and especially qualitative constraints that are induced from operator's execution traces. The induced constraints define the generalized operator's trajectory.

- Experimental evaluation of the proposed decomposition in several dynamic domains. It is shown that the induced generalized operator's trajectories provide a good insight into the operator's control skill and give rise to goal-directed and robust controllers.

- Development of an algorithm for learning of qualitative constraint trees from numerical examples. The algorithm partitions the attribute space to find qualitative functional constraints for each subspace and can handle noisy data.

- Applying the developed algorithm to induce qualitative models of human control skill. Such qualitative strategies define spaces for controller optimization and can be transformed into operational controllers. A method for combining domain knowledge of the control task with the induced qualitative strategy is given. It is shown that using simple facts about the control task improves the success of the resulting controllers.

# Chapter 2

# Learning Problem and Algorithms

This chapter first describes the machine learning problem for skill reconstruction and gives some notes on terminology. Section 2.3 describes dynamic domains and the simulators used in the experiments. These domains are controlling a container crane, a double pendulum acrobot, and bicycle riding. Section 2.4 briefly describes some standard machine learning algorithms used later in the thesis.

## 2.1 Learning Problem

The goal of skill reconstruction is twofold: to induce a successful controller and to achieve better understanding of the human operator's skill. A *successful controller* is a controller that carries out a given control task. In the experimental domains in this thesis a measure of controller's success is also the time required to accomplish the task. Among two successful controllers, the controller that requires less time to accomplish the task is considered to be more successful.

To achieve better understanding of the human operator's skill we are interested in learning symbolic and especially qualitative models of human control skills. As the learning examples, the logged data from skilled, human operators, i.e. operator's execution traces, are used. We will illustrate the learning problem with an example of the container crane dynamic system. This dynamic domain is later described in more detail (Section 2.3.1).

Container cranes are used in ports for load transportation. They are controlled by crane operators, who have acquired their skill through years of experience. Typically, such a control skill is subconscious and hard to rationally reconstruct through introspection (Urbančič, 1994; Bratko et al., 1998). So, the idea is to use their execution traces for automatic skill reconstruction by machine learning. An *execution trace* is just a sequence of the system's states and the corresponding operator's control actions, sampled at a certain frequency. At every time step, the values of all state and control variables are logged to a file. In the case of the container crane, control variables are the forces that the operator uses to control the system. System variables describe a state of the system, for example the position of the crane at a particular time point. Table 2.1 gives an example of such an execution trace from the container crane domain.

In the original approach to behavioural cloning such an execution trace is used to induce a tree predicting an action to take in a particular state. Every state-action pair is one learning example. The induced tree, called a clone, can be used to interpret the operator's skill or as

Table 2.1: An initial segment of a crane execution trace: $F_X$ and $F_L$ are the operators control actions, i.e. forces that the crane operator applies to the system at the particular times. The other variables describe the state of the crane.

| $X$ | $\dot{X}$ | $\Phi$ | $\dot{\Phi}$ | $L$ | $\dot{L}$ | $F_X$ | $F_L$ |
|---|---|---|---|---|---|---|---|
| 0 | 0 | 0 | 0 | 20 | 0 | 0 | 0 |
| 0 | 0 | 0 | 0 | 20 | 0 | 2500 | 0 |
| 0 | 0.01 | -0.01 | -0.02 | 20 | 0 | 6000 | 0 |
| 0.02 | 0.10 | -0.07 | -0.27 | 20 | 0 | 10000 | 0 |
| 0.12 | 0.31 | -0.32 | -0.85 | 20 | 0.01 | 14500 | 0 |
| 0.35 | 0.59 | -0.95 | -1.49 | 20.01 | 0.03 | 14500 | 0 |
| ... | ... | ... | ... | ... | ... | ... | ... |

an automatic controller of the system.

The described approach suffers from the problems pointed out in Section 1.2. In this thesis a different approach to the reconstruction of a human control skill from operator's traces is proposed. The learning problem is decomposed into two problems: (a) learning constraints on the operator's trajectories, called also the generalized operator's trajectory and (b) learning approximate models of system dynamics. To learn the system's dynamics we used $k$-Nearest Neighbour and locally weighted linear regression. The generalized operator's trajectory was learned by the machine learning program GoldHorn (Križman, 1998). These algorithms are briefly described in Section 2.4. To also learn the qualitative models of the operator's skill, algorithm QUIN that learns qualitative constraint trees from numerical examples is developed. This algorithm is described later in Chapter 5.

## 2.2 Notes on Terminology

Tanja Urbančič (1994) discusses different kinds of control knowledge:

> "With experience a human develops a behaviour that proved successful. He changes it only if forced by new circumstances. If we manage to describe it, we get a *strategy* and, by that, also an insight into a tacit *skill*. We gain also a certain degree of portability and the possibility to help those who are still acquiring the skill. A strategy could be described as "say how" (Michie, 1986)."

Similar to (Urbančič, 1994), we distinguish between different kinds of control knowledge:

- *Control skill* is operational, observable, but typically hard to describe "know how" at subsymbolic level.
- *Control strategy* is explicitly described, but not necessary fully operational "know how" at symbolic level.

In our approach to human skill reconstruction we consider in terms of symbolic constraints induced control strategy as a model of the operator's control skill. If we were able to fully reproduce a human control skill, we would get a controller that would control the system exactly as the human does. On the other hand, a control strategy does not necessarily define a unique controller, but rather a family of controllers. It can be described at different levels of abstraction. We use the terms *quantitative strategy* and *qualitative strategy* to distinguish between two different levels of abstraction. A quantitative strategy is given with precise, numerical values or numeric constraints on the parameters. A qualitative strategy may also be defined by qualitative constraints as used in qualitative reasoning. A qualitative strategy defines a family or a set of quantitative strategies.

Models at different levels of abstraction can be used to describe human control skill. What then is the most appropriate model of such a skill? And what is a valid model of human skill? At least for the first question, the answer depends on the goals of skill reconstruction. One important goal is to induce an understandable model of the underlying skill. Bain and Sammut (1999) note that one of the weaknesses of the original approach to behavioural cloning is that the generated rule sets are often very large and difficult to understand. Bratko and Urbančič (1999) give an example of the induced tree in the crane domain with 479 leaves for $F_X$ and 380 for $F_L$ and comment: "In spite of satisfactory behavioural reconstruction, such trees are unacceptable as models of human skill."

Modelling human skill at a higher level of abstraction, as a qualitative strategy for example, reduces the complexity of the induced rules and might therefore be easier to understand and reason about. Such qualitative models define an envelope within which the human operates. They only capture some properties of human control skill. As they do not completely replicate the operator's behaviour, are they still valid as models of human skills? We believe that skill reconstruction is not limited to the exact replication of the operator's behaviour. Several researches express a similar attitude, as explained in the following paragraphs.

In the context of expert systems Michie (1986) discusses the difficulties of the transfer of human "know how" to human "say how" by articulation and proposes the use of machine learning from human execution (or human "show how") to narrow the gap. He stresses the importance of interpretability of the induced rules, whether they make sense from the expert's point of view and whether they can be mentally checked. Establishing the correspondence between the clones induced from an operator's trace and the instructions provided by the operator, is also an important goal of human skill reconstruction (Urbančič and Bratko, 1994). As many human skills are sub-cognitive, such machine learning-based analysis can reveal the operators are occasionally not doing what they believe they are doing (Urbančič and Bratko, 1994). The utility of the induced rules is also important, that is whether they can be used to control the system in a similar way as the human does or help a novice in improving or acquiring the skill.

A relevant point also is that trees, induced by the original approach to behavioural cloning, do not completely replicate human control skill. One example is the *clean-up effect* phenomenon that is often observed with behavioural cloning. This phenomenon, i.e. that the clone sometimes controls the system in some way better, or more smoothly than the original human trace, was first reported by Michie, Bain and Hayes-Michie (1990) in pole-balancing domain and is in greater detail discussed in (Michie and Camacho, 1994). It was explained by filtering out small operator's inconsistencies and errors by the induction program. Michie (1995) also gives a negative effect of the clean-up effect and points out that "...learning from exemplary performance requires more than mindless imitation."

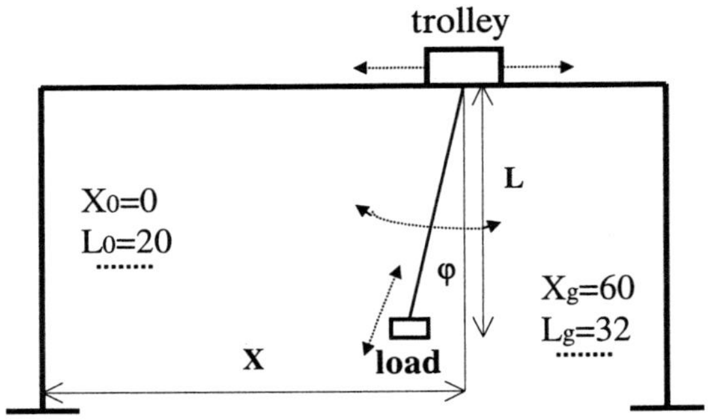


Figure 2.1: Container crane: the state of the system is specified by six variables: trolley position $X$ and its velocity $\dot{X}$, rope inclination angle $\Phi$ and its angular velocity $\dot{\Phi}$, rope length $L$ and its velocity $\dot{L}$. The system is controlled through force to the trolley $F_X$ and force in the direction of the rope $F_L$. The task is to transport the load from its start position to the goal region.

## 2.3 Experimental Dynamic Domains

### 2.3.1 Container Crane

**Container Crane and Related Work**

Container cranes are used in ports for load transportation. To transport a container (see Figure 2.1) from the shore to a target position on the ship, two operations are to be performed: positioning of the trolley, bringing it above the target load position ($X_g$), and rope operation, bringing the load to the desired height ($L_g$). The performance requirements include basic safety, stop-gap accuracy and the highest capacity possible. The last requirement means that the time for transportation is to be minimized. Consequently, the two operations are to be performed simultaneously. The most difficult aspect of the task is to control the swing of the rope. When the load is close to the goal position, the swing should ideally be zero.

As in previous experiments (Urbančič, 1994) in the crane domain, a crane simulator written by Marko Grobelnik (1994) is used. The parameters of the system (lengths, heights, masses, etc.) are the same as those of the real cranes in the Port of Koper in Slovenia. The state of the system is specified by six variables: trolley position $X$ and its velocity $\dot{X}$, rope inclination angle $\Phi$ and its angular velocity $\dot{\Phi}$, rope length $L$ and its velocity $\dot{L}$. Two control forces are applied to the system: force $F_X$ to the trolley in the horizontal direction and force $F_L$ in the direction of the rope. The simulator uses a fourth-order Runge-Kutta method based on the following equations of system's dynamics:

$$
\begin{aligned}
\ddot{X} =& \frac{1}{M}(F_X - F_L \sin(\Phi) + k(Mg + F_L \cos(\Phi))) \\
\ddot{L} =& g\cos(\Phi) + L\dot{\Phi}^2 - kg\sin(\Phi) + \frac{1}{m}(-F_L - F_{wz}\cos(\Phi) + F_{wx}\sin(\Phi)) \\
\ddot{\Phi} =& \frac{1}{2LmM}(F_Lkm - 2F_Xm\cos(\Phi) - 2kgmM\cos(\Phi) - \\
& - F_Lkm\cos(2\Phi) - 4mM\dot{\Phi}\dot{L} + 2F_{wz}M\sin(\Phi) - \\
& - 2gMm\sin(\Phi) + F_Lm\sin(2\Phi))
\end{aligned}
$$

Table 2.2: Parameters of the container crane simulator

| Parameter | Description | Value |
|---|---|---|
| $M$ | Trolley mass | 12800 kg |
| $m$ | Load mass | 17200 kg |
| $k$ | Trolley/rail friction coefficient | 0.02 |
| $F_{wx}$ | Force of air resistance ($x$ component) | 0 N |
| $F_{wz}$ | Force of air resistance ($z$ component) | 0 N |
| $g$ | Gravity acceleration | 9.81 $m/sec^2$ |

Table 2.3: Parameters of the crane control task. The task requires to transport the load from the start position to the goal region, i.e. a region around the goal state where each state variable is within the tolerance of 0.5% of its allowed range.

| Start pos. | Goal | Goal region | Allowed range |
|---|---|---|---|
| $X_0 = 0$ | $X_g = 60$ | $X \in [59.68, 60.32]$ | $X \in [0, 65]$ |
| $\dot{X}_0 = 0$ | $\dot{X}_g = 0$ | $\dot{X} \in [-0.02, 0.02]$ | $\dot{X} \in [-2.04, 2.04]$ |
| $\Phi_0 = 0$ | $\Phi_g = 0$ | $\Phi \in [-0.1, 0.1]$ | $\Phi \in [-10, 10]$ |
| $\dot{\Phi}_0 = 0$ | $\dot{\Phi}_g = 0$ | $\dot{\Phi} \in [-0.15, 0.15]$ | $\dot{\Phi} \in [-15, 15]$ |
| $L_0 = 20$ | $L_g = 32$ | $L \in [31.82, 32.18]$ | $L \in [3, 39]$ |
| $\dot{L}_0 = 0$ | $\dot{L}_g = 0$ | $\dot{L} \in [-0.01, 0.01]$ | $\dot{L} \in [-1.35, 1.35]$ |

The time step of the simulation is 0.02 seconds. Other parameters of the simulator are given in Table 2.2. The simulator offers two output modes ("picture" or "instruments"). "Picture" mode (see Figure 2.2) gives a graphical representation of the scene. In "instruments" mode, the state and the control variables are represented as bar graphs, imitating measuring instruments.

Previous experiments with the reconstruction of the human crane control skill include cloning with regression trees (Urbančič and Bratko, 1994; Urbančič, 1994). The main problem for regression trees was the swing control, i.e. large rope oscillations when the trolley approached its goal position. For this reason, slower, more conservative traces were found by far most useful for cloning with regression trees. Although successful clones were induced, the yield of cloning was low and the clones generally lacked robustness as reported also in (Bratko and Urbančič, 1999). An approach (Šuc and Bratko, 1997b; Šuc, 1998) that improves the robustness of clones exploits some results from control theory. Here clones take the form of linear-quadratic controllers with subgoals. This approach considers the system dynamics and practically enables identifying discrete subgoals. Some of our experiments, which are described later in the thesis, are also in (Šuc and Bratko, 1999a; Šuc and Bratko, 1999b).

### Experiments with Human Learning

We used experimental data from manually controlling the crane from a previous study (Urbančič, 1994). In that study, six students volunteered to learn to control a simulator. Their control task was to transport the load from a start position to a position near the goal state, called the goal region, and to maintain the load in the goal region for 8 seconds. The start position, the goal region and the allowed ranges of the state variables are defined in Table

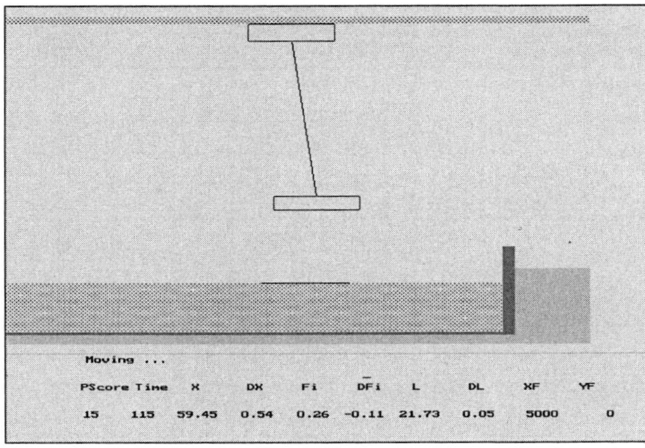


Figure 2.2: Crane simulator in "picture" mode: during the control graphical representation of the crane is displayed.

2.3. The time to complete the task is limited to 188 seconds. If the time exceeds 188 seconds, or if one of the state variables violates its allowed range, the control is not successful. The performance criterion is the time needed to conclude the task.

The subjects were not given any information about the system. They did not even know it was a crane. They were controlling the crane simulator in the "instruments" mode. Their task was to bring the bar graphs imitating the values of the state and control variables into a marked target region, where they should stay for a least 8 seconds. To enable the control of the force to the trolley $F_X$ from the keyboard, at each time step only two actions were possible: to increase the force or to decrease the force. $F_X$ was limited to the interval -15000 to 15000 N and the increase/decrease step was 500 N. The force to the rope $F_L$ was controlled similarly on the interval -2000 to 2000 N.

All subjects succeeded in accomplishing the task. However, remarkable individual differences were observed in the speed of control and the characteristics of the strategy they used. Some operators tended towards fast and less reliable operation, others were more conservative and slower in order to avoid large rope oscillations.

### 2.3.2 Acrobot

#### Acrobot and Related Work

The acrobot (Fig. 2.3) is an under-actuated (i.e. possessing fewer actuators than degrees of freedom) double pendulum. It consists of a two-link two-joint planar robot in a gravitational field. The torque control ($\tau$) can be applied to the elbow joint ($q2$), but the shoulder joint ($q1$) is free swinging. Any rotation at *q1* results entirely in dynamic coupling from the rest of the system. The dynamics of the acrobot is similar to a gymnast on a high bar where link 1 is analogous to the gymnast's hands, arms and torso, link 2 represents the legs, and joint $q2$ is the gymnast's waist. The acrobot's dynamics is given by:

$$d_{11}\ddot{q1} + d_{12}\ddot{q2} + h_1 + \Phi_1 = 0$$
$$d_{12}\ddot{q1} + d_{22}\ddot{q2} + h_2 + \Phi_2 = \tau$$

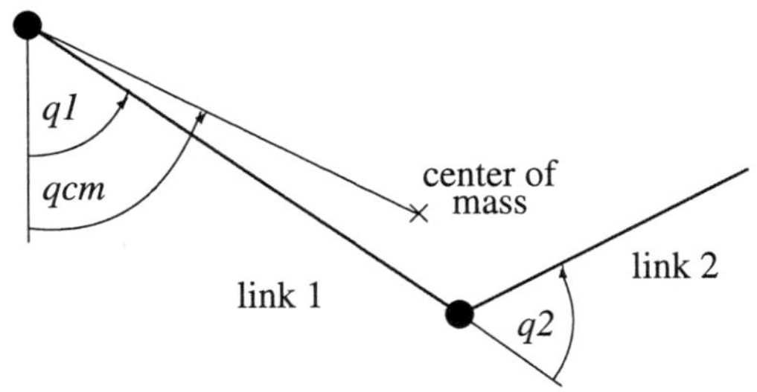


Figure 2.3: Acrobot, two link pendulum. Note that the angles $q1$ and $qcm$ are measured from the negative $y$ axis.

where:

$$\begin{aligned}
d_{11} &= m_1 l_{c1}^2 + m_2 l_1^2 + m_2 l_{c2}^2 + i_1 + i_2 + 2 m_2 l_1 l_{c2} \cos(q2) \\
d_{12} &= m_2 l_{c2}^2 + i_2 + m_2 l_1 l_{c2} \cos(q2) \\
d_{22} &= m_2 l_{c2}^2 + i_2 \\
h_1 &= -m_2 l_1 l_{c2} \sin(q2)(\dot{q2}^2 + 2\dot{q1}\dot{q2}) \\
h_2 &= \dot{q1}^2 m_2 l_1 l_{c2} \sin(q2) \\
\Phi_1 &= (m_1 l_{c1} + m_2 l_1) g \sin(q1) + m_2 l_{c2} g \sin(q1 + q2) \\
\Phi_2 &= m_2 l_{c2} g \sin(q1 + q2)
\end{aligned}$$

The parameters are explained in Table 2.4. The given dynamics is without friction. A simulator in LISP was kindly provided by Gerald DeJong (1995). As in his experiments, viscous friction was added to our implementation of the simulator. Viscous friction $f_1$ and $f_2$ at joints $q1$ and $q2$ makes the control task harder and affects the corresponding velocity by formula:

$$\begin{aligned}
\dot{q1}(t + dt) &= (1 - f_1)\dot{q1}(t) + dt\,\ddot{q1}(t) \\
\dot{q2}(t + dt) &= (1 - f_2)\dot{q2}(t) + dt\,\ddot{q2}(t)
\end{aligned}$$

The state of the acrobot is defined by angle $q1$ and its velocity and angle $q2$ and its velocity. Sometimes it is useful to consider also the center of mass angle, i.e. $qcm$, since it gives very important and easy-to-understand information about the acrobot position. As shown later in experiments with reconstruction of human swing-up skill (Section 4.3.1), the $qcm$ angle alone suffices for a simple, although not very efficient swing-up strategy. However, the $qcm$ angle can easily be expressed as a function of $q1$ and $q2$ and is not included in the acrobot state.

The acrobot is a test bed for nonlinear control theory. Some aspects of the acrobot are well understood. Designing a controller to balance an under-actuated double pendulum system vertically is studied in most beginners graduate control theory courses. The point corresponding to the upright position of the acrobot ($q1$=$\pi$, $q2$=0) is not the only unstable equilibrium point. There is a point of unstable equilibrium whenever the system's center of mass is directly above the $q1$ joint. The locus of these $q1$, $q2$ angle values forms a manifold. Maintaining balance while moving along this unstable manifold was an open problem until (Bortoff, 1992). One less well-understood problem is the swing-up control. Here, the task is to move the acrobot from its stable downward position to its unstable inverted (upright) position as fast as possible. A strategy to drive the controllable joint $q2$ so as to excite oscillation

Table 2.4: Parameters of the acrobot simulator

| Parameter | Description | Value |
|---|---|---|
| $l_1$ | Link 1 length | 0.12 m |
| $l_2$ | Link 2 length | 0.09 m |
| $l_{c1}, l_{c2}$ | Joint to mass center | $l_1/2, l_2/2$ |
| $i_1$ | Link 1 intertia | $m_1 {l_1}^2/12$ |
| $i_2$ | Link 2 intertia | $m_2 {l_2}^2/12$ |
| $m_1$ | Link 1 mass | 0.12 kg |
| $m_2$ | Link 2 mass | 0.09 kg |
| $g$ | Gravity acc. | 9.8 $m/sec^2$ |
| $dt$ | Simulator step | 0.003 sec |
| $f_1, f_2$ | Friction in joints $q1$,$q2$ | 0, 0.012 |
| $\tau$ | Torque at middle joint ($q2$) | $\lvert\tau\rvert \leq 0.4Nm$ |
| $q2_{max}$ | Maximum $q2$ deflection angle | 2.2 rad. |

of $q1$, must be found. The oscillation must grow until a point of the unstable equilibrium, i.e. when the system's center of mass is directly above the $q1$ joint.

The swing-up task is difficult for several reasons. First, any solution involves periodic driving. The controller must cycle between different control movements. Such periodic excitation often leads to uncontrollable behaviour. Indeed, in the acrobot system most of the periodic excitations of $q2$ lead quickly to a chaotic regime from which establishing control would be difficult or impossible. Second, the system dynamics is highly nonlinear. In addition, handling viscous friction which depends on the velocity is particularly difficult. With wide oscillations come high velocities, which result in high frictional losses from frictional damping. Viscous losses at each joint combine a nonlinear pattern due to the joint's kinetic energy together with the dependence on the position and motion of the other joint.

The difficulty of the swing-up task attracted the attention of many researchers. A control theory solution for the swing-up task was derived by Spong (1994). This strategy, called ATAN, chooses a control trajectory of $q2 = a_1 arctan(a_2 q1)$. The derivation of this solution requires the complete model and tuning of the controller parameters $a_1$ and $a_2$. A different approach which requires no model is reinforcement learning. Here, the policy is learned through trial and error on the simulator or a physical system. The disadvantage of this approach is in extensive use of the physical system and the lack of interpretability of the learned policy. Sutton (1996) successfully used Q-learning methods to learn a discrete action swing-up control policy. To reduce the amount of data required from the real robot, Boone (1997) used heuristic knowledge which helped the search in reinforcement learning. DeJong (1995) found an efficient swing-up strategy using an interesting idea, called Explanation Based Control. This approach makes extensive use of qualitative domain knowledge and requires one successful execution trace. Qualitative knowledge determines the space of possible control strategies, which is empirically searched. The control strategy with the simplest qualitative explanation that accounts for the entire given execution trace is selected. Some of the experiments described later in the thesis are also in (Šuc and Bratko, 2000).

### Experiments with Human Learning

As a part of our study experiments with humans learning to control the acrobot were carried out. Four subjects volunteered to learn the *swing-up strategy* on the simulator. Their control task was to reach the point where the system's center of mass is directly above the $q1$ joint ($qcm = \pm\pi$), starting from the stable downward position ($q1$=$\dot{q1}$=$q2$=$\dot{q2}$=0). Stabilization in the unstable inverted position was not required. The performance criterion is the time needed to conclude the task.

The simulator parameters were as used in (DeJong, 1995), with some minor changes enabling humans to control the system. This parameters were determined in initial experiments with the simulator. The maximum $q2$ deflection angle was set to $q2_{max}$=2.2 rad., since human operators found it very hard to learn a swing-up strategy with the original maximum $q2$ angle 1.5 rad. If $|q2|$ angle exceeded the maximum $q2$ deflection angle, the trial was unsuccessfully terminated. In order to enable torque control input from the keyboard, only two actions were possible at each time step: to increase the torque or to decrease the torque. The torque control was limited to the interval -0.4 to 0.4 Nm and the increase/decrease step was 0.08 Nm. Viscous friction was set to 0.0 at $q1$ joint and to 0.012 at $q2$ joint. Another experiment with viscous friction set to 0.002 and 0.012 in the corresponding joints was made, but none of the subjects succeeded in completing the swing-up task under that condition. The rest of the parameters are defined in Table 2.4. The simulator also provides a graphical representation of the scene, i.e. the swinging acrobot, besides the values of the state and control variables.

Each subject performed 150 trials. Due to the relatively fast dynamics of the system, for human learning the simulator was slowed down by setting the time between two consecutive states of the simulator (and the control time step) to 0.05 sec. of real "operator's time". This gives the slow down of approximately 18 and was chosen as the slow down that human operators felt comfortable with. A trial was successful if the swing-up task was completed in 16000 steps (800 sec. of "operator's time") and $|q2|$ never exceeded the maximum $q2$ deflection angle. The first constraint was not critical since skilled operators usually complete the task much faster. Respecting the $q2_{max}$ limit was harder, at least at the beginning of the learning process.

To develop their own control strategies, the subjects were not allowed to observe other subjects learning. Every execution trace of each subject was recorded. All the four subjects succeeded in accomplishing the task. However, remarkable individual differences were observed regarding the speed of learning as well as the effectiveness of their swing-up strategies. Typically, a successful trial lasted about 2000 steps, that is 6 seconds of simulated time or 100 seconds of the subjects' real time. However, skilled subjects could complete the task in about 1500 steps and required 5 to 6 swings.

### A Simple Swing-up Strategy

Every possible solution to the swing-up task involves periodic oscillations of $q1$ such that the amplitude of $q1$ increases in consecutive cycles. This means that the solution requires several cycles, where each cycle consists of two *swing-up* and two *swing-down* parts. Lets define:

- *swing-up* part when $sign(q1) = sign(\dot{q1})$. During swing-up $|q1|$ is increasing and $\left|\dot{q1}\right|$ is decreasing.

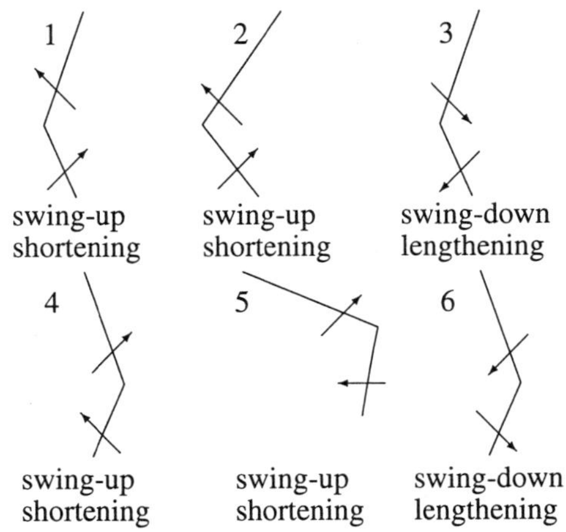


Figure 2.4: One cycle of a simple swing-up strategy: during swing-up part of the cycle $|q2|$ is increased (shortening of the acrobot), and during swing-down $|q2|$ is decreased (lengthening of the acrobot), so $|q2|$ reaches its local maximum at the end of the swing-up (when $|q1|$ reaches its local maximum) and $q2$ crosses zero when the acrobot swings down through the vertical.

- *swing-down* part when $sign(q1) \neq sign(\dot{q1})$. During swing-down $|q1|$ is decreasing and $|\dot{q1}|$ is first increasing and then, just before the end of swing-down ($|q1|$ approaches zero) $|\dot{q1}|$ is decreasing.

At the beginning of their experimenting with the acrobot most of the subjects used the simple strategy illustrated in Fig 2.4. The idea of this strategy is to drive the second link so as to inject the energy in phase with the swinging of the first link, somewhat similar to a child pumping a swing: when link 1 ($q1$ angle) rotates counterclockwise, $q2$ is controlled to move clockwise, and when link 1 rotates clockwise $q2$ is controlled to move counterclockwise. During the swing-up part of the cycle, $|q2|$ is increased (*shortening* of the acrobot), and during the swing-down, $|q2|$ is decreased (*lengthening* of the acrobot).

### 2.3.3 Bicycle Riding

#### Bike Domain and Related Work

Here the control task that requires maintaining balance on a bike and driving it to the goal is described.

The bike's speed is considered to be constant. The state of the bike is described by six variables: the tilt $\omega$ angle of the bicycle from vertical and its velocity $\dot{\omega}$, the angle $\theta$ between the front wheel direction and the bicycle direction due to the deflection of the handlebars and its velocity $\dot{\theta}$, the distance $D_g$ from the goal and the angle $\Psi_g$ of the bike's frame relative to the goal. The system is controlled by two actions: torque $\tau$ applied to the handlebars and displacement $d$ of the bicyclist from the center of mass. The control task is to maintain balance on the bike and drive from the start position to the goal position with the constant speed of 10 km/h. We used the simulator that was kindly provided by Randløv and Alstrøm (1998). The details of the simulator are given at the end of this section.

Maintaining balance on the bike is a very hard task and requires a proper adjustment of the front wheel direction by applying a torque to the handlebars and/or a proper displacement of the center of mass. The requirement to approach the goal makes the task even harder, as the balancing and the goal-aiming are to be performed simultaneously.

In previous experiments (Randløv and Alstrøm, 1998) with reinforcement learning in this domain a technique called shaping was used to speed up the learning. Here, the agent was first given the task to learn to balance the bike and after that the task to reach the goal while using the skill of balancing. Another approach used in this domain is called Iterated Phantom Induction (Brodie and DeJong, 1999). This approach uses qualitative domain knowledge to bias the learning and requires less experimentation with the system.

### Experiments with Human Learning

The learning examples for the reconstruction of bicycle riding skill were execution traces from an experiment where six students learned to manually control the bike simulator and complete the control task. The simulator, apart from values of the state and the control variables, also provides a graphical representation of the scene, i.e. the bicycle in motion. Since the dynamics of the system are very fast, the simulator was slowed down by factor of 6. This was the slow down that human operators felt comfortable with.

Operators were required to balance the bike and drive it to the goal that was at 100 m from the start position, with the initial frame orientation of the bike $-\pi/2$ from the goal direction. More precisely: their control task was to drive a bike from the start position ($\omega = \dot{\omega} = \theta = \dot{\theta} = 0, D_g = 100, \Psi_g = -\pi/2$) to the 5 m radius of the goal ($D_g < 5$). A trial was accepted as successful if the bike reached the 5 m radius of the goal within 100 seconds and did not fall over on the way. The measure of the operator's success was the time needed to complete the task.

While developing their own control strategies, the subjects were not allowed to observe other subjects learning. All six subjects succeeded in accomplishing the task. Each of the operators performed (a) 100 trials by using torque control only (no weight-shifting), plus (b) 50 trials using both displacement and torque control. First they learned to balance the bike using only torque control. When they were allowed to use also the displacement control, they usually used it only for the initial turn towards the goal, but this did not significantly improve their performance.

The operators found the task very difficult, since maintaining the balance requires precise adjustment of the handlebars and/or displacement of the center of mass. Losing concentration for just a moment results in a large tilt angle velocity $\dot{\omega}$ and the bike falls. The requirement to approach the goal makes the task even harder, since balancing and goal-aiming must be performed simultaneously. For this reason all the operators needed more than 25 unsuccessful trials before completing the control task for the first time. The first successful trials of an operator were usually far from optimal: they turned to the goal slowly and sometimes even drove in circles before reaching the goal position. Afterwards the operators improved their performance, so most of the operators were able to complete the task in less than 45 seconds of simulated time, that is 270 seconds of subjects' real time. The most successful operator needed 37 seconds of simulated time to complete the task, which is very close to the minimal possible time.

Table 2.5: Parameters of the bike simulator

| Parameter | Description | Value |
|---|---|---|
| $c$ | Horizontal distance between mass center (CM) and point where front wheel touches ground | 66 cm |
| $d_{\text{CM}}$ | Vertical distance between mass centers of bicycle and cyclist | 30 cm |
| $h$ | Distance of mass center from ground | 94 cm |
| $l$ | Distance between front and back tyre at the point where they touch the ground | 111 cm |
| $M_c$ | Mass of bicycle | 15 kg |
| $M_d$ | Mass of tyre | 1.7 kg |
| $M_p$ | Mass of cyclist | 60 kg |
| $r$ | Tyre radius | 34 cm |
| $\dot{\sigma}$ | Tyre angular velocity | $\dot{\sigma} = \frac{v}{r}$ |
| $v$ | Velocity of bicycle | 10 km/h |

### Details of the Bike Simulator

We used the simulator with the parameters as in (Randløv and Alstrøm, 1998). The simulator is controlled by two actions: the torque $\tau$ applied to the handlebars and the displacement $d$ of the bicyclist from the center of mass. Possible values of torque $\tau$ are -2,-1,0,1,2 N and for $d$ -2, 0, or 2 centimeters. Simulator time step $dt$ is 0.01 sec. Other parameters are given in Table 2.5.

The bicycle system is modelled by non-linear differential equations. The equations do not model the system exactly since some second order cross effects were ignored during the derivation. The equations that were kindly provided by the authors (Randløv and Alstrøm, 1998) of the simulator are:

$$\varphi \stackrel{\text{def}}{=} \omega + \arctan\left(\frac{d}{h}\right)$$

$$\ddot{\omega} = \frac{1}{I_{\text{bicycle and cyclist}}} \times \left(Mhg\sin\varphi - \cos\varphi\left(I_{dc}\,\dot{\sigma}\dot{\theta} + \text{sign}(\theta)\cdot v^2\left(\frac{M_d\,r}{r_f} + \frac{M_d\,r}{r_b} + \frac{Mh}{r_{\text{CM}}}\right)\right)\right)$$

$$\ddot{\theta} = \frac{\tau - I_{dv}\,\dot{\sigma}\,\dot{\omega}}{I_{dl}}$$

Here CM denotes the center of bicycle mass and the cyclist together and $d$ the displacement of the cyclist from CM perpendicular to the plane of the bicycle. $\varphi$ is the angle of the tilt of the mass center. The front and back tyres follow different paths in a curve with different radii. Front tyre follows a longer path. The radius for front and back tyres, the radius of the

CM and estimated moments of inertia are:

$$r_f = \frac{l}{|\cos(\frac{\pi}{2} - \theta)|} = \frac{l}{|\sin\theta|}$$
$$r_b = l\left|\tan\left(\frac{\pi}{2} - \theta\right)\right| = \frac{l}{|\tan\theta|}$$
$$r_{\text{CM}} = \left((l-c)^2 + \frac{l^2}{(\tan\theta)^2}\right)^{\frac{1}{2}}$$
$$I_{\text{bicycle and cyclist}} = \frac{13}{3}M_c h^2 + M_p\,(h + d_{\text{CM}})^2$$
$$I_{dc} = M_d\,r^2, I_{dv} = \frac{3}{2}M_d\,r^2, I_{dl} = \frac{1}{2}M_d\,r^2$$

The positions of the front and the back tyres are computed as:

$$\begin{pmatrix} x_f \\ y_f \end{pmatrix}_{(t+1)} = \begin{pmatrix} x_f \\ y_f \end{pmatrix}_{(t)} + v\,dt \begin{pmatrix} -\sin(\psi + \theta + \text{sign}(\psi + \theta)\arcsin(\frac{v\,dt}{2r_f})) \\ \cos(\psi + \theta + \text{sign}(\psi + \theta)\arcsin(\frac{v\,dt}{2r_f})) \end{pmatrix}$$
$$\begin{pmatrix} x_b \\ y_b \end{pmatrix}_{(t+1)} = \begin{pmatrix} x_b \\ y_b \end{pmatrix}_{(t)} + v\,dt \begin{pmatrix} -\sin(\psi + \text{sign}(\psi)\arcsin(\frac{v\,dt}{2r_b})) \\ \cos(\psi + \text{sign}(\psi)\arcsin(\frac{v\,dt}{2r_b})) \end{pmatrix}$$

## 2.4 Learning Algorithms

### 2.4.1 Instance-based Learning and $k$-Nearest Neighbour

In contrast to learning methods that construct a general, explicit description of the class function when training examples are provided, instance-based learning methods (Stanfill and Waltz, 1986; Aha et al., 1991; Stanfill and Waltz, 1986; Mitchell, 1997) simply store the training examples. Each time a new query instance is encountered its relationship to previously stored examples is examined in order to assign the class value for the new instance. Instance-based learning includes nearest neighbour (Duda and Hart, 1973) and locally weighted regression methods (Cleveland, 1979; Atkeson et al., 1997a; Atkeson et al., 1997b) that assume that instances can be represented as points in a Euclidean space. Instance-based methods are sometimes referred to as "lazy" learning methods because they delay processing until the next instance must be classified. A key advantage of this kind of delayed, or lazy learning is that instead of estimating the class function once for the entire instance space, these methods can estimate it locally and differently for each new instance to be classified.

The most basic instance based method is $k$-Nearest Neighbour algorithm. The class value of the new query instance is calculated as the mean value of the $k$ nearest neighbours if the class is continuous-valued, or as the majority value, if the class is discrete-valued. The nearest neighbours of an instance are defined in terms of the standard Euclidean distance.

### 2.4.2 Locally Weighted Linear Regression

The nearest neighbour approach can be thought of as approximating the class function at a single query point. Locally weighted regression is the generalization of this approach and constructs an explicit approximation of the class function over a local region surrounding the query point. Locally weighted linear regression (LWR) uses distance-weighted training examples to form this local linear approximation to the class function.

There is a broad range of alternative methods for distance weighting the training examples and tuning fit parameters of the local model (Atkeson et al., 1997a). Our implementation (see (Šuc, 1998) for details) uses Gaussian kernel weighting. Let $d_i$ denote the Euclidean distance of the $i$-th training example from the query point. Then the weight of the $i$-th example is:

$$w_i = e^{\frac{-d_i}{2K^2}} \tag{2.1}$$

Here $K$ denotes Gaussian kernel that is a user-defined parameter. The weighted training examples are then used for linear regression to regress the coefficients of the linear function. By changing $K$, more (small $K$) or less importance is assigned to the distance and therefore more local (small $K$) or more global linear model is learned. To set $K$ automatically, the optimization of the mean squared cross-validation error can be used.

Locally weighted linear regression was successfully used in a number of reinforcement learning applications (Schaal and Atkenson, 1994; Atkeson et al., 1997b). In this thesis we use LWR to approximate the system's dynamics from the execution traces. Basically, any nonlinear function approximator could be used for this purpose, but we prefer LWR because:

- it enables *incremental learning*: when the system is controlled, each current state is simply stored, and this experience can be used to predict the system's behaviour in nearby points in the state space. If we expect the system's dynamics to change in time, only recent experience is considered in the prediction phase.

- it provides a local *linear model* of the system's dynamics near the current state. A linear model of system dynamics enables that the linear function is "turned around" to express the control action achieving the desired next state. Consider for example a simple nonlinear dynamic system, where a control force $F$ is applied to an object, and the desired velocity $\dot{X}_{des}$ of the object is known. We can use LWR to learn (approximate) local linear model of system dynamic, i.e. $\dot{X}(t+dt) = aX(t) + b\dot{X}(t) + cF(t)$ and then compute the action aiming to achieve the desired velocity by: $F = (\dot{X}_{des} - aX(t) - b\dot{X}(t))/c$.

### 2.4.3 GoldHorn

GoldHorn (Križman, 1998; Križman et al., 1995) is a machine learning system intended to discover empirical laws that govern the behaviour of dynamical systems. Given the behaviour of the system, i.e. a sampled execution trace, GoldHorn attempts to find a set of ordinary differential equations that describe the dynamics of the system. GoldHorn does not simply fit the parameters of equations of given forms, but it also constructs new forms of equations. To do this, GoldHorn first introduces new terms by repeatedly applying operators, such as multiplication, to the state variables and their time derivatives. Then, given the set of all the terms, including the original variables and the newly constructed terms, differential equations are generated from these terms using linear regression. The significance of the equation is judged by two measures: the multiple correlation coefficient and the normalized deviation. It should be noted, that as a result, GoldHorn not only finds differential equations of some predefined forms, but also does *structural* synthesis of new forms of equations.

GoldHorn is an extension of the program Lagrange (Džeroski and Todorovski, 1993). The improvements with respect to Lagrange effectively deal with noisy data. This makes GoldHorn applicable to the modelling of real world systems from imperfect measured data. Other equation discovery systems as BACON (Langley et al., 1987), ABACUS (Falkenhainer and Michalski, 1986) and FAHRENHEIT (Koehn and Zytkow, 1986) discover empirical numerical laws from data, but Lagrange and GoldHorn are unique since they are intended for dynamical systems.

Here GoldHorn is not used to learn the dynamics of the system, but to induce constraints among state variables when the system is controlled by an experienced operator. In this way, GoldHorn is used to induce the operator's control strategy.

# Chapter 3

# Problem Decomposition for Behavioural Cloning

In the original approach to behavioural cloning of the human operator's skill, a controller is usually induced directly as a classifier from the system's states into actions. Typically, the induced control rule is a decision or regression tree representing a function $u = u(x)$, where $x$ is the state vector of the dynamic system, and $u$ is the control action to be performed in $x$.

This chapter proposes and explores a decomposition of the cloning problem into two learning problems: the learning of operator's control trajectories and the learning of the system's dynamics. This decomposition tackles two problems that were observed with the original approach to behavioural cloning and commented in Section 1.2:

- Typically, induced clones are brittle with respect to small changes in the control task.
- The clone induction process typically has low yield: the proportion of successful controllers among all the induced clones is low, typically well below 50% (excluding wrapper-based approaches commented in Section 1.2).

Section 3.1 describes the proposed problem decomposition for behavioural cloning and defines indirect controllers as opposed to direct controllers learned by the original approach to behavioural cloning. Section 3.2 presents an experiment where the robustness of learned direct and indirect controllers against the learning error is compared and gives a characterization of the learner's error, that is a plausible explanation of why this decomposition approach has empirically proved to be usually superior to *direct* cloning. Section 3.3 empirically compares the cloning yield of direct and indirect controllers in the domain of container cranes. The chapter concludes with possible advantages of indirect controllers and some comments on related work.

## 3.1 Problem Decomposition and Indirect Controllers

Here we describe the decomposition of the behavioural cloning problem and define indirect controllers as opposed to direct controllers learned by the original approach to behavioural cloning. We use the following notation.

Let $X$ denote the set of system's states and $\mathcal{U}$ the set of possible actions. $x$ denotes the system's state vector, $x_k$ the system's state at time point $k$, and $x_g$ the goal state. $u$ denotes the

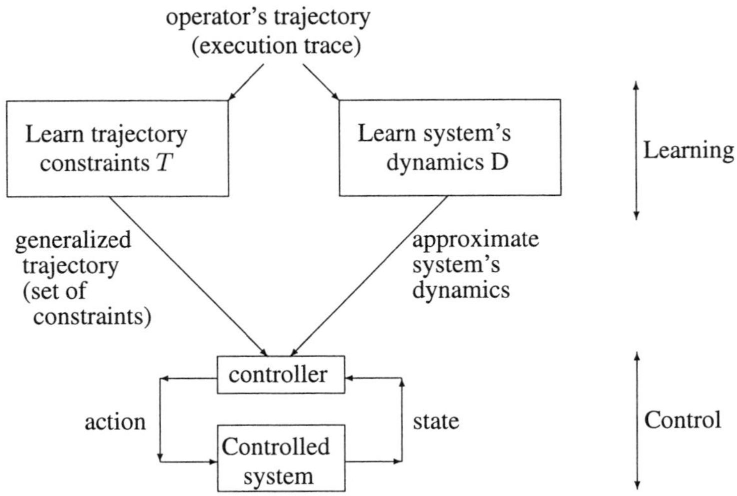


Figure 3.1: The indirect controller can be induced from an operator's control trace by generalizing the operator's trajectory and learning the system dynamic's model. Both together are used to control the system.

control action vector. $y$ denotes a vector of the chosen state variables (also called *dependent* variables) and $x_y$ the system's state vector with variables $y$ excluded. An execution trace is a sequence of system's states and corresponding actions, sampled at a certain frequency, i.e. $(x_k, u_k), k = 0 \ldots N$, where $N$ is the last time point that was recorded. A *trajectory* is the corresponding sequence of system's states, i.e. $(x_k), k = 0 \ldots N$.

### 3.1.1 Learning an Indirect Controller

The construction of a controller by induction from the operator's traces consists of three stages (see Figure 3.1):

1. Learn constraints $T$ on the operator's trajectories, which can be formally stated as a mapping $T$:
$$T : X \times X \mapsto \{true, false\} \tag{3.1}$$
Constraints $T$ define what will be called a *generalized operator's trajectory*. The trajectory constraints are usually represented as a function $t$, expressing a chosen state variable $y$ (dependent variable) as a function of other variables in the system's state, i.e. $y^T = t(x_y)$. If the system is controlled by more than one control variable, it might be necessary to express more than one dependent variable as a function of the system's state (then $y$ is a vector). A suitable chosen dependent variable can be a state variable that is "most directly" affected by the available actions according to the human notion of causality in the system. For example, for the pole-cart system (Michie and Chambers, 1968) the generalized trajectory could be represented as a function mapping position of the cart, angle of the pole and angular velocity of the pole to velocity of the cart, i.e. $\dot{x} = t(x, \theta, \dot{\theta})$. Another criterion for selecting a suitable dependent variable

can be comprehensibility, i.e. we prefer to choose dependent variables which are close to human thinking about controlling a particular dynamic system. This, of course, requires the human's domain knowledge and his/her notion of the causality in the system as discussed in Chapter 7. For example, when controlling a car, the driver would probably prefer the constraints on car's speed than the constraints on the accelerator pedal position or the engine revolutions.

2. Learn an appropriate model $D$ of the system's dynamics. $D$ can, for example, be a function that maps the system's states and actions to acceleration, eg.: $\ddot{x} = f(x, \dot{x}, u)$. More generally the system's dynamics model can be learned as a function mapping system states and actions into states at the next time step, eg.: $x_{k+1} = D(x_k, u_k)$.

3. Once trajectory constraints $T$ and dynamics model $D$ have been induced, they are used to construct a controller, which works as follows:

   (a) Given a current system state $x_k$ at time point $k$, $\Delta$ = diff$(x_k, T)$ denotes the deviation between the state and the generalized operator's trajectory. Usually the deviation is defined simply as the difference between the value of the chosen dependent variable defined by $T$ and its current value.

   (b) Use the system's dynamics model $D$ to determine an action $u_k$ which reduces the deviation $\Delta$ in time. For discrete time $k$, an appropriate action $u_k$ in the current state $x_k$ can be:

   $$u_k = \arg\min_{u \in \mathcal{U}} \text{diff}(x_{k+1}, T) \tag{3.2}$$

   where $x_{k+1} = D(x_k, u)$ and $\mathcal{U}$ is the set of possible actions. We call this an *indirect controller* as opposed to a *direct controller* that computes action directly as $u_k = u(x_k)$ (see Figure 3.2).

### 3.1.2 Some Technical Notes

Computing the action $u_k$ which reduces the deviation between the state and the generalized operator's trajectory according to Equation 3.2 requires minimization of $\Delta$ = diff$(D(x_k, u), T)$ over all possible actions $u$. This might be computationally expensive when the set of possible actions is large. One solution to this problem is to learn a local linear model of the system dynamics and use its inverse to directly compute the action, aiming to achieve the desired next state on the trajectory. This method was already mentioned in Section 2.4.2 and is given below.

Given the generalized trajectory and the current state $x_k$, the desired values of the dependent trajectory variables can be computed as $y^T_{k+1} = t(x_{y_k})$. The action aiming to achieve the desired next state on the trajectory is computed as follows. First, learn a local (near current state $x_k$) linear model of system dynamics that expresses the values of dependent variables at the next time step given the current state and the action, i.e. learn matrices $A_k$, $B_k$ and vector $c$ from $y_{k+1} = A_k x_k + B_k u_k + c_k$. Then the action $u_k$ which aims to achieve the desired next state on the generalized trajectory is $u_k = {B_k}^{-1}(y^T_{k+1} - A_k x_k - c_k)$.

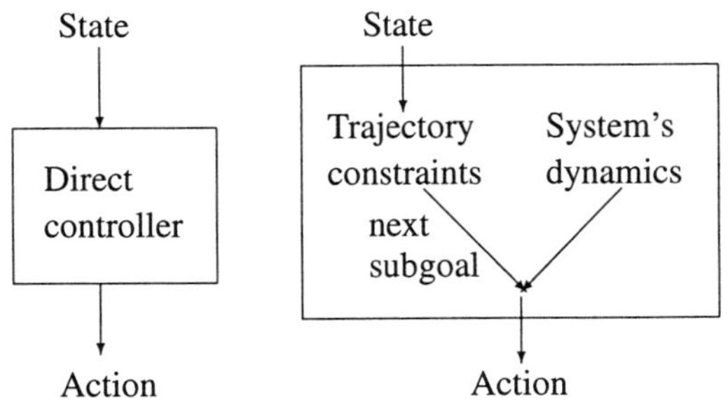


Figure 3.2: Direct and indirect controller: on the left is an direct controller, and on the right is an indirect controller. A direct controller uses a direct mapping from states to actions, whereas the indirect controller uses trajectory constraints and a model of the system dynamics to take the action towards the next subgoal.

### 3.1.3 Direct and Indirect Controllers

A controller observes the current state $x$ and takes a corresponding action $u$, i.e, it is a function mapping states to control actions: $X \mapsto \mathcal{U}$. This controller function can take different forms. In the case of a linear controller ((Bertsekas, 1987; Franklin et al., 1986)), for example) it is the product of the error $(x - x_g)$ in state $x$ and a matrix $K$, while in the case of the original approach to behavioral cloning, it is a regression or a decision tree. The classic control theory approach to controller design usually involves the modelling of the system's dynamics, whereas the original approach to behavioral cloning does not use the system's dynamics model and learns a direct mapping from states to actions.

To make a distinction between the controllers that do and do not use a system's dynamics model and to compare the learnability of such controllers, we define two kinds of controllers. A *direct controller* corresponds to a direct mapping from the observed state to the control action. An *indirect controller* uses the system's dynamics model and some intermediate concepts to achieve the goal-directedness of the controller and therefore corresponds to an indirect mapping from states to actions. Approaches like BOXES (Michie and Chambers, 1968; Sammut, 1994), ASE/ACE, (Barto et al., 1983) and the original approach to behavioral cloning learn direct controllers. A controller using the generalized trajectory (Šuc and Bratko, 1997a; Šuc and Bratko, 1999a; Šuc and Bratko, 1999b; Šuc and Bratko, 2000) is an indirect controller. The classical linear controller can also be viewed as an indirect controller. These controllers use the system's dynamics model and intermediate concepts in the form of goal $x_g$ (linear controller) or generalized trajectory. Behavioural cloning approaches like GRAIL (Bain and Sammut, 1999), CHURPs (Stirling, 1995) and Incremental Correction model (Camacho, 1998; Camacho, 2000) also make use of a problem decomposition and a kind of indirect controllers. These approaches are discussed in Section 3.5.

One could argue that a tree, induced with the original approach to behavioural cloning, is an indirect mapping too, because it maps state $x$ to a generalized state represented by a leaf in the tree and then maps this generalized state to action $u$. If we are lucky, the generalized state is an understandable intermediate concept, perhaps a region in a state space where the operator takes similar actions. However, the trees induced with the original approach to behavioural cloning in general lack conceptual structure that would clearly capture the causal relations in the domain and the goal structure of the control strategy (Urbančič and Bratko, 1994; Bratko et al., 1995; Michie, 1995; Bratko and Urbančič, 1999). This of course assumes that these elements are present in the operator's control strategy.

In the case of a generalized trajectory, the controller is a composite of two functions: the trajectory function $t : x_y \mapsto y^T$ and the inverse dynamics $d : x \times y^T \mapsto u$. The trajectory function maps the state with excluded dependent variables $x_y$ to the desired vector of dependent variables $y^T$ and therefore gives constraints on the desired next state. So, typically the desired next state is not completely determined; some of the state variables are omitted. Another possibility, described in the next chapter, is to define the trajectory function using qualitative constraints. The inverse dynamics function maps the current state and the desired vector of dependent variables to an action which aims to achieve the desired next state on the trajectory.

## 3.2 Comparing Direct and Indirect Controllers

### 3.2.1 Robustness against Learning Error

Here we study and compare the performance of direct and indirect controllers learned from an operator's execution trace. We assume deterministic system's dynamics $f$ and discrete time, i.e. $x_{k+1} = f(x_k, u_k)$, where index $k$ indicates successive sample times $t_0, t_0 + dt, \ldots,$ $t_0 + k\,dt$. $x$ denotes the state vector and $u$ the control action vector.

Both, direct and indirect controllers, are learned from an operator's execution trace $((x_k^{op}, u_k^{op}), k = 0 \ldots N)$. A direct controller is induced by learning the actions. We call an action of a direct controller a *generalized action*. An indirect controller is induced by generalizing the trajectory $((x_k^{op}), k = 0 \ldots N)$ and the model of system's dynamics. In the following examples we show that a direct controller is prone to the departure from the trajectory and that this departure is likely to happen due to the learning error of the generalized action.

The operator takes the action $u_k^{op}$ in the current state $x_k^{op}$ that achieves the desired next state $x_{k+1}^{op}$. The learned controller mimics the operator and takes action $u_k$ that achieves the desired next state with some error $\epsilon_{k+1}$: $x_{k+1} = f(x_k, u_k) = x_{k+1}^{op} + \epsilon_{k+1}$.

A property of many nonlinear dynamic systems is that an action applied "far" from the trajectory tends to cause the system to move in a different direction than when the same action is applied near the trajectory. Intuitively a direct controller is not likely to be successful in such a system. Since the generalized action $u_k^D$ is learned with some error, and these errors are often propagated to the next state, the controller's trajectory is likely to diverge "far" from the operator's trajectory. A direct controller is not likely to return on the operator's trajectory from such a region.

As opposed to a direct controller, an indirect controller's action $u_k^I$ uses the learned dynamics at the current state $x_k = x_k^T + \epsilon_k$ when aiming to approach the next state on the trajectory $x_{k+1}^T$. Even if the current state $x_k$ is "far" from the operator's trajectory, the controller explicitly aims at decreasing the error in the next step. Since the system's dynamics model can be updated online, the indirect controller can approach the operator's trajectory even from the states not seen in the operator's trace.

### 3.2.2 Experiment

Here we compare the robustness of direct and indirect controllers with respect to the learning error on an example of a simple dynamic system control. We wanted to make our experiment independent of the particular learning technique, so we modelled the "induced" predictors simply by taking the correct value and corrupting it with noise. No learning was actually done with direct controllers. They used "correct" actions (operator's actions) corrupted with noise. Indirect controllers used the generalized operator's trajectory corrupted with noise. In addition, indirect controllers also had to cope with the error in the system's dynamics model that was induced with a simple kind of instance-based learning.

We consider a deterministic, discrete time dynamic system with the following dynamics:

$$\begin{aligned} x_{k+1} &= x_k + \dot{x}_k \\ \dot{x}_{k+1} &= x_{k+1} + u_k x_{k+1}^2 + u_k \end{aligned} \tag{3.3}$$

A similar system was used in (Atkeson et al., 1997b) as an example of a nonlinear system that is easily controlled with an optimal linear controller (i.e. with a linear quadratic controller) in the region near zero, but hard to control elsewhere in the state space.

The task is to control the system to reach and maintain the goal position $x_g = 0.5$ from the start state ($x_0 = \dot{x}_0 = 0$). The criterion of the controller's success is the controller's error $c$. Here we define $c$ as the sum of absolute errors in $x$ for the first $N = 30$ steps: $c = \sum_{k=0}^{k=N} |x_k - x_g|$. If $|x_k|$ exceeds 100 the error is set to its maximum $c = 100$. Since $x_0 = 0$ and $x_1 = 0$, $c \geq 1$.

We suppose that the operator uses a very simple control rule that does the task with $c = 1.04$:

$$u_k^{op} = \begin{cases} 0.5, \text{if } x_k \leq 0 \text{ and } \dot{x}_k = 0 \\ -0.4, \text{if } x_k > 0 \text{ and } |x_g - x_k| \leq 0.001 \\ -0.4 - 0.001\ sign(x_g - x_k), \text{otherwise} \end{cases} \tag{3.4}$$

making the system move approximately along the trajectory $((x_0, \dot{x}_0), \cdots, (x_{k+1}^{op}, \dot{x}_{k+1}^{op}), \cdots)$, where:

$$\begin{aligned} x_{k+1}^{op} &= x_k + \dot{x}_k \\ \dot{x}_{k+1}^{op} &= \begin{cases} (x_g - x_k) - \dot{x}_k, \text{if } x_k = 0 \\ 0.09(x_g - x_k), \text{otherwise} \end{cases} \end{aligned} \tag{3.5}$$

A generalized trajectory is specified by constraints between the current and the next state. Since the control $u_k$ directly affects only $\dot{x}_{k+1}$ (and not $x_{k+1}$), an appropriate generalized trajectory can be learned as a mapping from ($x_k$, $\dot{x}_k$) to the desired $\dot{x}_{k+1}^T$ and the indirect controller's action $u_k^I$ can be computed from the learned system's dynamics. In our experiments with indirect controllers we used a very simple kind of instance-based learning to learn (during the control) a local model of the system's dynamics near the current state $(x_k, \dot{x}_k)$: 10 examples $((x_j, \dot{x}_j), u_j, (x_{j+1}, \dot{x}_{j+1}))$ with the smallest quadratic distance $(x_j - x_k)^2 + (\dot{x}_j - \dot{x}_k)^2$ are used for linear regression to compute the parameters $a$,$b$,$c$,$d$ of the local linear model $\dot{x}_{k+1} = ax_k + b\dot{x}_k + cu_k + d$ used at the point $(x_k, \dot{x}_k)$. The points used for instance-based learning are continuously updated during the control, i.e. starting with the set $\{\ x_k^{op}, \dot{x}_k{}^{op}, u_k^{op}\ \}$ and adding new points during the control. The indirect controller's action

Table 3.1: Robustness of direct and indirect controllers against the learning error: $c^D_{reactive}$ denotes the error of "reactive" direct controllers with the learning errors modelled as Gaussian noise with standard deviation $\sigma$. $c^D$ and $c^I$ denote the controller error for the direct and the indirect controllers with the learning errors modelled as Gaussian noise with standard deviation $\sigma$. $c^D*$ and $c^I*$ denote the controller's error where the predictors' error was modelled by *biased* Gaussian noise. The results are averages of 40 runs.

| learning | direct controllers | | | indir. controllers | |
|---|---|---|---|---|---|
| error $\sigma$ | $c^D_{reactive}$ | $c^D$ | $c^D*$ | $c^I$ | $c^I*$ |
| 0.0 | 1.0 | 1.0 | 1.0 | 1.0 | 1.0 |
| $10^{-2}$ | 51.5 | 1.7 | 3.2 | 1.6 | 2.6 |
| 0.1 | 75.5 | 9.4 | 58.2 | 8.7 | 18.1 |
| 0.2 | 94.4 | 19.2 | > 100 | 20.2 | 39.2 |
| 0.5 | > 100 | 86.6 | > 100 | 83.8 | 95.0 |

$u^I_k$, aiming to achieve the desired next state on the generalized trajectory is then computed as $u^I_k = (\dot{x}^T_{k+1} - ax_k - b\dot{x}_k - d)/c$.

From the execution trace $((x^{op}_k, \dot{x}_k{}^{op}, u^{op}_k),\ k = 0, \ldots, N)$, where Rule 3.4 was used, the generalized operator's action $u^D_k$ and the generalized operator's trajectory $\dot{x}^T_k$ can be learned with the learning errors $\epsilon^D$ and $\epsilon^T$, i.e. prediction errors due to the generalization by the learning system. So, the predicted values of the induced predictors are:

$$u^D_k = u^{op}_k + \epsilon^D_k$$
$$\dot{x}^T_k = \dot{x}^{op}_k + \epsilon^T_k$$

As noted at the beginning of this section, no learning was actually done with direct controllers. They used "correct" actions currupted with noise modelling the learning error. At first we experimented with direct controllers that use the operator's action $u^{op}_k$ (Rule 3.4) currupted with learning error $\epsilon^D_k$ modeled as randomly generated zero mean Gaussian noise with various standard deviations $\sigma$. The actions were computed as $u^D_k = u^{op}_k + \epsilon^D_k$. The results given in Table 3.1 and Figure 3.3 show that such direct controllers are very brittle with respect to the learning error. The reason is that the Rule 3.4 is purely reactive. These "reactive" direct controllers have considerably higher controller errors (see Figure 3.3) than direct controllers we used afterwards.

In the rest of the experiments we used a better control rule for $u^{op}_k$. It uses the system dynamics (Equation 3.3) to achieve the desired $\dot{x}^{op}_{k+1}$ from Rule 3.5. Therefore the direct controller's action is:

$$u^D_k = \frac{\dot{x}^{op}_{k+1} - x^{op}_{k+1}}{1 + (x^{op}_{k+1})^2} + \epsilon^D_k \tag{3.6}$$

In this way, direct controllers were given the advantage of exact system model, while indirect controllers used a very simple method to approximate the system dynamics.

To compare the robustness of direct and indirect controllers with respect to the learning error we performed a set of experiments, where we modeled different error distributions of the learning system and measured the performance of both controllers. We experimented with two prediction error models. First the error was randomly generated as zero mean Gaussian noise with various standard deviations $\sigma$. The second model is the same as Gaussian noise,

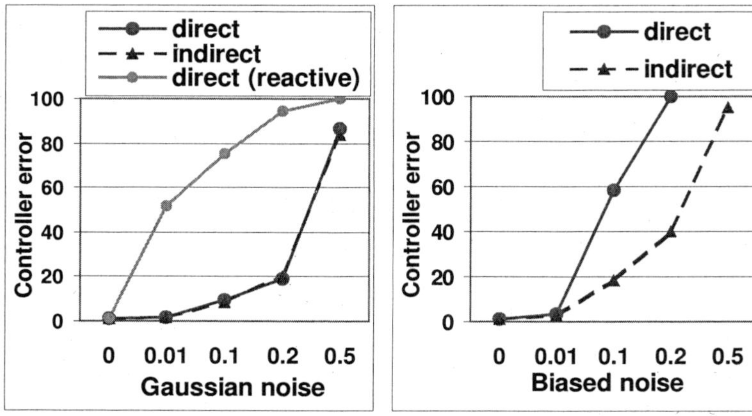


Figure 3.3: Robustness of direct and indirect controllers against the learning error. $y$-axis denotes controller error. On $x$-axis is $\sigma$ of the learning error, modeled as Gaussian (left graph) and *biased* Gaussian noise (right graph). Solid line corresponds to direct controllers, dashed line to indirect controllers, and pale solid line to "reactive" direct controllers (Rule 3.4). The results are averages of 40 runs.

but with all the errors (in the same control trace) in the same direction, that is all errors in the same control trace are either positive or negative, with random absolute value. We call this *biased* Gaussian noise. Biased Gaussian noise is obtained from Gaussian noise simply by adjusting the sign (making the sign uniform for the whole control trace). Note that biased Gaussian noise thus has the same $\sigma$, but of course no longer has zero mean.

The rationale for defining our experiment in this way is as follows. We wanted to make our experiment independent of the particular learning technique, so we modeled the "induced" predictors simply by taking the correct value and corrupting it with noise. Gaussian noise as error model seems to be an obvious choice. The second model, biased Gaussian noise is, however, not quite so obvious. We designed this error model to account for a frequent property of induced continuous predictors: namely, that small changes in the attribute values usually do not cause large changes in the predicted class values. Consequently, the errors at neighbour points have the same sign.

Figure 3.3 gives average errors 40 of direct and indirect controllers, where the learning error was modeled as biased Gaussian and Gaussian noise with $\sigma$=0, 0.01, 0.1, 0.2, 0.5. The results show that Gaussian noise in the respective predictors affect both direct and indirect controllers in terms of their control cost to a similar extent. Biased Gaussian noise is more damaging to both controllers than zero mean Gaussian noise with the same $\sigma$. However, the direct controllers are affected much more than the indirect controllers. This happens, surprisingly, in spite of the fact that in addition to the error in generalized trajectory, the indirect controller also has to cope with the error in the induced dynamics model. This result provides a plausible explanation why indirect controllers have proved to be much more successful in typical dynamic domains such as the crane and the acrobot. These experiments are described in the following section and in Chapter 4.

We believe that this superiority of the indirect controller arises from their goal-directedness (generalized operator's trajectory can be seen as the continuous subgoal) and their use of the system's dynamics model, which is learned online and thus enables them to behave well also in the yet unseen regions of the state space. The awareness of the system's dynamics is important since it enables the controllers to be the goal-directed. However, there are different ways how to consider the system's dynamics. Other behavioural cloning approaches (see Section 3.5) like CHURPs (Stirling, 1995), GRAIL (Bain and Sammut, 1999) and Incremental Correction model (Camacho, 1998; Camacho, 2000) also consider the dynamics of the system.

## 3.3 Experiments in the Crane Domain

In the previous section we compared robustness (against the learning error) of direct and indirect controllers on a simple dynamic system by modelling different error distributions of the learning system. In this section we compare the success of direct and indirect controllers on a complex dynamic system by using real learners. Execution traces of experienced operators were used as the learning examples for $k$-Nearest Neighbour learning method to induce direct and indirect controllers for a dynamic system of container cranes. This domain was described in Section 2.3.1. Besides comparing the success of controllers, the motivation for the experiment is also to study the effects of (a) the size of the learning set and (b) mixing different control styles on learning direct and indirect controllers.

We chose $k$-Nearest Neighbour as the learning method for its simplicity and since it requires to set a relatively small number of parameters. Of course we could choose another learning method, but as illustrated for example in (Michie, 1995), using a different learning method usually does not result in such improvements as using an appropriate problem decomposition.

First we give the details of learning direct and indirect controllers from execution traces and describe the learning examples for learning direct and indirect controllers. In Section 3.3.3 we give experimental results.

### 3.3.1 Learning Direct and Indirect Controllers

To learn direct and indirect controllers from execution traces we used $k$-Nearest Neighbour. Direct controllers used $k$-Nearest Neighbour to directly learn the control actions $F_X$ and $F_L$ to apply to the system. Indirect controllers learned the desired trajectory of trolley and rope velocity, i.e. $(\dot{X}_{des}, \dot{L}_{des})$ trajectory, by $k$-Nearest Neighbour and approximate model of system dynamics by locally weighted linear regression (LWR). This local linear model is used to compute the control actions $F_X$ and $F_L$ aiming to achieve the desired next state on the trajectory.

The dependent trajectory variables $\dot{X}_{des}$ and $\dot{L}_{des}$ were chosen according to the notion of the causality in the system, namely that these are the state variable that are the most directly affected by control actions $F_X$ and $F_L$.

To select $k$ nearest neighbours we used weighted Euclidean distance measuring the similarity of the current state and learning example states. To help learning some domain knowledge was incorporated into this similarity measure. Let $d(v)$ denote the normalized difference of state variable $v$ in the current and the learning example state, i.e. the difference divided by the maximal range of the variable $v$. Let $w(v)$ denote the weight giving the relative importance of difference in the state variable $v$. When learning direct controller action $F_X$

or indirect controller $\dot{X}_{des}$ the nearest neighbours are the $k$ states minimizing the distance $\sum_{v\in\{X,\dot{X},\Phi,\dot{\Phi}\}} w(v) \times d^2(v)$. When learning direct controller action $F_L$ or indirect controller $\dot{L}_{des}$ the nearest neighbours minimize $\sum_{v\in\{X,L,\dot{L}\}} w(v) \times d^2(v)$. The weights were set to $w(X)$=$w(L)$=1, $w(\Phi)$=$w(\dot{X})$=$w(\dot{L})$=0.5, $w(\dot{\Phi})$=0.2, according to the domain knowledge, namely that the distances in some state variables are more important than the distances in other variables.

Note that the same domain knowledge was used for both controllers. When experimenting with all weights $w(v)$ set to one, the relations between percentages of successful direct controllers and percentages of successful indirect controllers were similar.

The controllers work as follows. The direct controller at each time step observes the current state and takes actions $F_X$ and $F_L$ by averaging corresponding actions of $k$ nearest neighbours. The indirect controller observes the current state and first learns the desired trolley velocity $\dot{X}_{des}(t)$ and the desired rope velocity $\dot{L}_{des}(t)$ by averaging corresponding $\dot{X}(t+dt)$ and $\dot{L}(t+dt)$ of $k$ nearest neighbours. Afterwards it learns a model of system dynamics near the current state by LWR and uses this approximate model to select actions $F_X$ and $F_L$ aiming to achieve the desired $\dot{X}_{des}(t)$ and $\dot{L}_{des}(t)$. Precisely, LWR learns local linear model near the current state, i.e. $\dot{X}$ and $\dot{L}$ at next time step as a linear function of the state variables and the corresponding force:

$$\begin{aligned}
\dot{X}(t{+}dt) &= k_1X(t) + k_2\dot{X}(t) + k_3\Phi(t) + k_4\dot{\Phi}(t) + k_5L(t) + k_6\dot{L}(t) + k_7F_X(t) \\
\dot{L}(t{+}dt) &= l_1X(t) + l_2\dot{X}(t) + l_3\Phi(t) + l_4\dot{\Phi}(t) + l_5L(t) + l_6\dot{L}(t) + l_7F_L(t)
\end{aligned}$$

Actions $F_X$ and $F_L$, aiming to achieve the desired $\dot{X}_{des}(t)$ and $\dot{L}_{des}(t)$ are then computed from the linear model:

$$\begin{aligned}
F_X &= (\dot{X}_{des} - k_1X(t) - k_2\dot{X}(t) - k_3\Phi(t) - k_4\dot{\Phi}(t) - k_5L(t) - k_6\dot{L}(t))/k_7 \\
F_L &= (\dot{L}_{des} - l_1X(t) - l_2\dot{X}(t) - l_3\Phi(t) - l_4\dot{\Phi}(t) - l_5L(t) - l_6\dot{L}(t))/l_7
\end{aligned}$$

It is important to note that the model of system dynamics used by indirect controllers is *approximate* and learned during the control. To get a notion of the error of the learned system dynamics model 3 traces of the same operator were used for learning the system dynamics by LWR and mean absolute errors of $\dot{X}(t+dt)$ and $\dot{L}(t+dt)$ on other three traces of the same operator were measured. Mean absolute errors were respectively 0.009 and 0.002, that is about 0.2% of the variable range or 3% of the maximal variable change in one time step. Better results can be achieved with different parameters of LWR, but we did not try to improve the learning of the system dynamics. In this way, indirect controllers also had to cope with the error of approximate system dynamics model.

### 3.3.2 Learning Examples

Successful execution traces were used as the learning examples to induce direct and indirect controllers. To study also the effects of (a) the size of the learning set and (b) mixing different control styles we combined execution traces into *trace sets*. A trace set is simply a union of different traces that are used as the learning examples to induce a controller. Traces of 3 operators, performing different control strategies, were combined into trace sets as follows.

Let $o$ denote the number of different operators, whose traces were combined into a trace set and $t$ the number of traces of the same operator. We chose five successful traces of every

Table 3.2: The description of the learning trace sets: $o$ denotes the number of different operators, whose traces were combined into a trace set and $t$ denotes the number of traces of the same operator. The number of traces in a trace set is $T = o \times t$. Five successful traces of every operator were combined into trace sets. We used $N_{dif}$=15 corresponding trace sets consisting of every single trace ($o$=1,$t$=1). For other ($o$,$t$) combinations we generated $N_{dif}$=24 corresponding trace sets.

| $o$ operators | $t$ traces | $T$ $T = o \times t$ | $N_{dif}$ | Description |
|---|---|---|---|---|
| 1 | 1 | 1 | 15 | 1 operator, 1 trace |
| 1 | 2 | 2 | 24 | 1 operator, 2 traces |
| 1 | 3 | 3 | 24 | 1 operator, 3 traces |
| 2 | 3 | 6 | 24 | 2 operators, 3 traces of each |
| 3 | 1 | 3 | 24 | 3 operators, 1 trace of each |
| 3 | 3 | 9 | 24 | 3 operators, 3 traces of each |

operator and combined them into trace sets consisting of $t$=1,2,3 traces of the same operator ($o$=1,$t$=1,2,3), trace sets consisting of $t$=1,3 traces of each operator ($o$=3,$t$=1,3) and trace sets consisting of 3 traces of each pair of operators ($o$=2,$t$=3). We used all possible trace sets consisting of a single trace ($o$=1,$t$=1). For other ($o$,$t$) combinations we generated 24 corresponding trace sets. This gives $15 + 24 \times 5$ different trace sets. The number of traces in a trace set is $T = o \times t$. Therefore, we have 15 different traces sets for $T$=1, 24 trace sets for $T$=2,6,9 and 48 different trace set for $T$=3 (combinations ($o$=1,$t$=3) and ($o$=3,$t$=1)). Table 3.2 gives the description of the trace sets used.

The traces were sampled with sample time $dt = 0.2$ sec. Since a successful trace typically lasts for about 90 seconds, one trace gives about 450 learning examples.

### 3.3.3 Experimental Results

The evaluation criterion in the experiment was the percentage of successful controllers among all the learned controllers. From every trace set a direct and an indirect controller was learned with $k$=3,5,10,20,50,100,200.

Experiments showed that the original crane control task is very hard for learning direct controllers by $k$-Nearest Neighbour. Only one of the direct controllers successfully finished the original crane control task. This is the direct controller learned from 3 traces of the same operator with $k$=5. Therefore, the maximal success of direct controllers is 4.1%. Indirect controllers performed much better (see Figure 3.4). The success rate is often over 20%. The most successful (50%) are indirect controllers learned from 9 traces with $k$=50. As expected, the success of indirect controllers increases with the increasing trace set size $T$. Also, with the increasing $T$, bigger $k$ do better.

Since direct controllers failed the original crane control task, we also conducted experiments with a modified crane control task that requires less accuracy in achieving the goal region. The tolerance in achieving the goal region was modified from 0.5% in the original task to 2.5% in the modified control task. Figure 3.5 gives the results of learning direct and indirect controllers with the modified crane control task. Again, indirect controllers do much better. The dependence of controllers' success on $k$, i.e. the number of nearest neighbours is

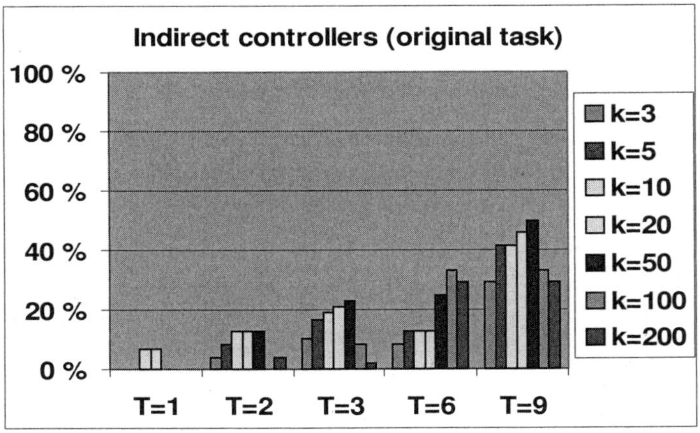


Figure 3.4: Learning indirect controllers for the original crane control task: the $x$-axis is marked with the number of traces $T$ used for learning. The $y$-axis gives the percentage of successful controllers. Different columns correspond to different $k$ used for $k$-Nearest Neighbour. From every trace set a direct and an indirect controller was learned with each $k$. Therefore, the results for each $k$ are the average of 15 controllers for trace set size $T$=1, 48 controllers for $T$=3 and 24 controllers for $T$=2,6,9. The missing columns correspond to 0%.

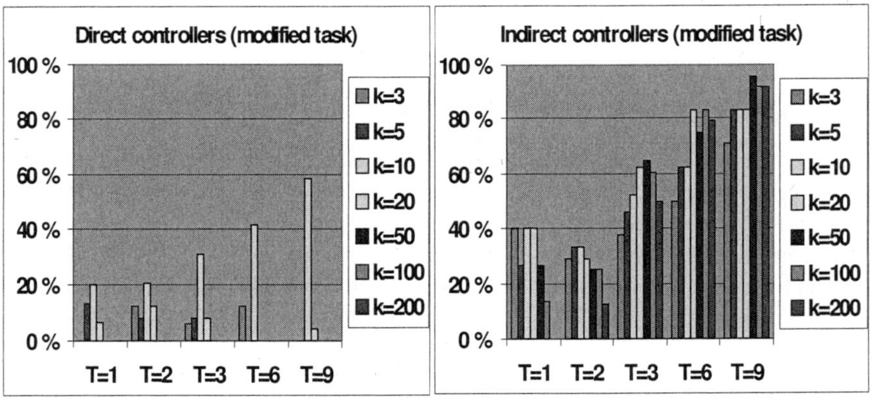


Figure 3.5: Learning indirect and direct controllers for the modified crane control task: the $x$-axis is marked with the number of traces $T$ used for learning. The $y$-axis gives the percentage of successful controllers. Different columns correspond to different $k$ used for $k$-Nearest Neighbour. The results for each $k$ are the average of 15 controllers for trace set size $T$=1, 48 controllers for $T$=3 and 24 controllers for $T$=2,6,9. The missing columns correspond to 0%.

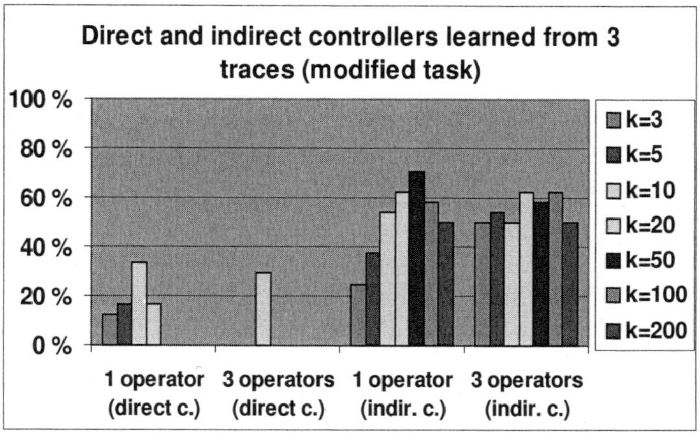


Figure 3.6: Learning indirect and direct controllers from 3 traces (modified crane control task): The $y$-axis gives the percentage of successful direct and indirect controllers learned from 3 traces of the same operator and 3 traces of 3 different operators. Different columns correspond to different $k$ used for $k$-Nearest Neighbour and are the average of 24 controllers.

very interesting. Indirect controllers are successful with a wide range of different $k$, whereas direct controllers are much more sensitive to changes in $k$: $k$=10 yields far better results than any other $k$. This might lead to a speculation that learning indirect controllers is less sensitive to the parameters of the learning method.

To study the effect of mixing different control styles we also compared the success of direct and indirect controllers that were learned from 3 traces of the same operator and 3 traces of 3 different operators. The results given in Figure 3.6 are quite surprising. Combining different control styles does not affect indirect controllers much. Direct controllers learned from different operators are less successful than when traces of the same operator are used. This observation is interesting for the following reason. As noted in (Michie, 1993) "The weak point of any technique based on learning from an expert exemplar is that training samples tend to be distributed over only narrow regions of the total state space." The induced (direct) controller is not given enough examples to recover when the system is not in the desired state. Considering the fact that mixing different control styles gives more *variety in examples* and thus better coverage of the state space, these results might contradict the intuition that by better coverage of the state space the success of direct controllers should increase. Similar observation, i.e. that mixing traces from different operators is damaging for learning direct controllers with decision trees, was reported in (Bratko et al., 1998).

To summarize the observations in the experiment of this section:

- The yield of learning indirect controllers is considerably better than when learning direct controllers.

- Indirect controllers are successful with a wide range of different $k$, whereas direct controllers are much more sensitive to changes in $k$. This might lead to a speculation that learning indirect controllers is less sensitive to the parameters of the learning method.

- Combining different control styles is damaging for direct controllers, whereas it does not affect indirect controllers much. Considering the fact that mixing different control styles gives more *variety in examples* and thus better coverage of the state space, these results might contradict the intuition that by better coverage of the state space the success of direct controllers should increase.

## 3.4 Possible Advantages of Learning Indirect Controllers

Here we would like to comment on the plausible advantages of learning indirect controllers. It should be noted that the discussion in this section is based on the experiments in this thesis in which particular learning methods were used. If a different method was used, the result would possibly be different. However, on the basis of substantial experience, we argue that differences in particular learning methods would not qualitatively affect the results.

We believe that most of these advantages are the consequences of the decomposition of learning an indirect controller into two natural subproblems: learning what to do (the trajectory) and learning how to do it (the system's dynamics). Experiments presented in this chapter and later in the thesis suggest that indirect controllers are robust with respect to the learning error of the system's dynamics model. The only difficult task is to learn the generalized trajectory. As confirmed by experimental results, the system's dynamics can be approximated sufficiently well by instance based methods of learning from execution traces and can be also updated online during the control.

The experiments indicate the following possible advantages of learning indirect controllers:

1. An indirect controller is less prone to the departure from the desired trajectory. Due to the operator's inconsistency (mistakes, taking different actions in similar situations) and due to the learning errors the learned trajectory or the action is not perfect. When taking an incorrect action, the system often departs from the desired trajectory and arrives at yet unseen region of the state space. Indirect controller is goal directed and considers the system's dynamics to take an action back towards the trajectory. A direct controller has no clue how to get back to the known region of the state space and often just gets stuck in the unknown region.

2. An indirect controller is more robust against changes in the system's dynamics and small changes in the task. The robustness against change in the system's dynamics is ensured by updating the system's dynamics model during the control. The robustness against small changes in the task (like changing the system start state) follows from better robustness of the indirect controller against the departure from the desired trajectory. This assumes, of course, that the learned generalized trajectory is still appropriate for the changed system's dynamics or the changed task.

3. Generalizing the trajectory is often easier than generalizing the actions. In many systems different actions can result in similar effect on the system behavior. When selecting the action to take, an experienced operator often keeps in mind the desired effect only. (S)he sometimes uses different actions to achieve the same effect in similar situations. These different actions in similar states are effectively noise for the learning system when learning the mapping from the system's states to the actions, but are not noise when learning the mapping from the system's states to the desired effects, that

is generalized trajectory. For example, when controlling a crane, the operator can use quite different actions to slow down the trolley: a sequence of oscillating actions around $F_X$=2000 can result in similar behavior as a sequence of actions $F_X$=2000.

4. Learning an indirect controller seems usually to be an easier task than learning a direct controller. When learning an indirect controller, the only difficult task is to learn the generalized trajectory, that is what the operator is doing. In the case of a direct controller the task is to learn what the operator is doing and how (s)he does it. This seems to be a more difficult learning task.

In addition, in the experiments, indirect controllers had other advantages that are further discussed later in the thesis:

1. The learned trajectory is easier to understand than the actions, which usually involve knowledge of the system's dynamics. The desired trajectory usually captures the knowledge of performing the given task in a more compact way. For example, when driving a car on a parking space from one point to the other, it is easier to understand the (x,y) trajectory than the sequence of actions like pushing the accelerator pedal, turning the wheel for 30 degrees and straightening it after, pushing the breaks, turning the wheel again,...
2. Knowledge learned by generalized symbolic trajectory can easily be adopted to new tasks. For example, in the crane domain, the trajectory for attaining the goal position at X=60 can easily be adopted to goal position at X=80.

## 3.5 Related Work

Probably the first work on goal-structured learning of control was done by Leech (Leech, 1986) by an application of machine learning to controlling a chemical process. He used decision trees in two steps. First to predict the value of measured variables as a function of control variables, and second to predict the yield of the plant based on the values of the measured variables. These rules are then combined to control the system so as to maximize the yield of the plant.

Leech's work is different from our approach in the following. We decompose the cloning problem into learning the dynamics and learning the generalized trajectory. On the other hand, he decomposed the learning of the "dynamics" model (yield as a function of control variables) into two steps. But there is in fact no dynamics in his model because the time behaviour in the system state transition is not modelled at all. Only the stationary state behaviours are handled in his model. Learning the trajectory was in his case not necessary, since the desired "trajectory" ("maximize yield") was given in advance. His work demonstrated that problem decomposition might help the learner. The values of measured variables (from induced trees) that maximize the yield can also be seen as goals and might improve the comprehensibility of the controller. In contrast to the generalized trajectory, which can be seen as a continuously changing subgoal, his goals do not change and his strategy therefore does not involve time at all. In this respect Leech's cloning problem and the corresponding decomposition is simpler than that in our case.

A similar idea of decomposing the cloning problem into two learning problems appears in GRAIL (Bain and Sammut, 1999). GRAIL stands for Goal-directed Reactive Abduction from Inductive Learning. It is a methodology by Bain and Sammut intended as an extension of the behavioural cloning methodology to handle goals and overcome robustness problems. A GRAIL controller comprises an effects level and a goals level, both based on rule-sets induced from the operator's traces. Abduction is used to learn effects of control actions and decision trees to learn subgoals. Preliminary results with GRAIL in the piloting domain show improvements in terms of reduced sizes of rule-sets, however, the authors do not report on experiments to investigate benefits regarding the success of controllers constructed in this way. The authors stress that the extension of GRAIL to use structured theories and background knowledge via first-order learning is most likely to further improve the performance of behavioural cloning.

An interesting approach, called CHURPs (Stirling, 1995), showed that considering a system's dynamics model and subgoals improves the robustness of controllers. CHURPs stands for Compressed Heuristic Universal Reaction Planners and explicitly separates the control skills in planning and actuation phases. It assumes that a complex task can be broken down into a finite set of sub-tasks, each with clearly defined subgoal. The actuation component of the human skill is modelled as simple regulatory feedback control (tracking) between a control input and its assigned subgoal. The planning component enables CHURPs to allocate alternative, surrogate controls if a preferred control input is unavailable or has failed. One similarity to our approach is that it enables to take into account the human knowledge of the relationships between the controls and the effects of the controls. CHURPs uses the influence matrix for this purpose. In contrast to our approach, CHURPs uses machine learning only to get the rules telling which combination of control variables to use in order to change a given state variable. Subgoals are not learned, but are assumed to be a part of the task specification.

A decomposition of the cloning problem also appears in Incremental Correction model (Camacho, 1998; Camacho, 2000). This approach uses two kinds of goals. Achievable goals have a well-defined set of start and final states in the state space. Homeostatic goals are achieved continuously and do not terminate. The Incremental Correction model has two levels of decisions and is based on conjecture that humans use an "incremental correction" strategy to do real-time control. The higher level decides whether the situation is acceptable or not. An acceptable situation is defined by a small deviation from homeostatic goal values and requires no change in the control values. A non acceptable situation initiates the low level that takes a sequence of control variable corrections until the situation is back to an acceptable status. Parameterized decision trees are used to learn how to achieve or maintain the goals. The values of achievable and homeostatic goals are randomly generated from a range of admissible values. A wrapper (John et al., 1994) is used as a method to tune the parameters of machine learning algorithm. Still, producing a successful controller requires testing of different clones (induced with different parameters) on the simulator. The main problem is the lack of correlation between the error measured on the independent test set and the performance of the resulting controller. This work demonstrates that using subgoals might improve the robustness of clones. It also reduces complexity of the clones in terms of the size of the induced rules. However, the approach requires a definition of the control task plan as a temporal sequence of stages and is limited to homeostatic goals that do not change in time. Our generalized trajectory can be seen as a continuously changing homeostatic goal.

Another approach that deals with learning to control dynamic systems is reinforcement learning (Sutton and Barto, 1998; Kaelbling et al., 1996). Unlike behavioural cloning, reinforcement learning learns control from scratch, through trial-and-error. The idea is to use dynamic programming, to learn the value function, or $Q$-function estimating how promising (in achieving the goal) each state or action is. In the most common approach direct controllers are learned. To reduce experimentation with the dynamic system, model-based reinforcement learning methods (Sutton, 1990) also use learned model of system dynamics and therefore effectively learn indirect controllers. However, reinforcement learning usually does not use human operator's control skill and is not concerned with the comprehensibility of the learned strategy.

Our experiments with the decomposition of the cloning problem began with skill reconstruction in the acrobot domain (Šuc and Bratko, 1997a).

# Chapter 4

# Symbolic and Qualitative Skill Reconstruction

In the previous chapter we proposed a problem decomposition for behavioural cloning and showed that learning indirect controllers might improve the yield of cloning and the robustness of the resulting clones. Another important goal of behavioural cloning is comprehensibility of a learned controller and to achieve better understanding of the operator's subconscious skill. For this reason we model operator's control skill by *symbolic and qualitative constraints*. In this chapter, we follow the idea of the cloning problem decomposition and use symbolic machine learning to generalize the operator's trajectories. Furthermore, qualitative abstraction is used to derive the operator's qualitative strategy that is comprehensible and offers the possibility to improve the operator's control strategy.

First a general description of the approach is given. In Sections 4.2 and 4.3 experiments in the crane and acrobot domains are presented. Finally, some points of interest are discussed and conclusions drawn.

## 4.1 Overview

We follow the idea of the cloning problem decomposition from the previous chapter and use symbolic machine learning to generalize operator's trajectories. Learning the controller is decomposed into generalizing the operator's trajectory by equation learning program GoldHorn and learning the system's dynamics by locally weighted linear regression. When generalizing a given operator's trajectory, we do not learn the trajectory in time, but rather induce constraints among the state variables in the operator's execution trace (or traces). Usually, the induced constraints involve only some of the state variables and have the form of differential equations. Both, the generalized trajectory and the nonlinear model of system dynamics form an indirect controller as explained in Section 4.1.1.

Generalized trajectory in the form of equations induced by GoldHorn defines a *quantitative strategy*. The strategy is "quantitative" (as opposed to *qualitative*) in the sense that it defines the precise *numerical* values of the chosen trajectory variables as a function of other state variables. Such quantitative strategy can be used to control the system. Also, it provides some insight into the operator's control strategy. Since equations are often hard to interpret, we use qualitative abstraction to transform the induced quantitative strategy into a *qualitative strategy*. This qualitative strategy is expressed in terms of qualitative constraints as described

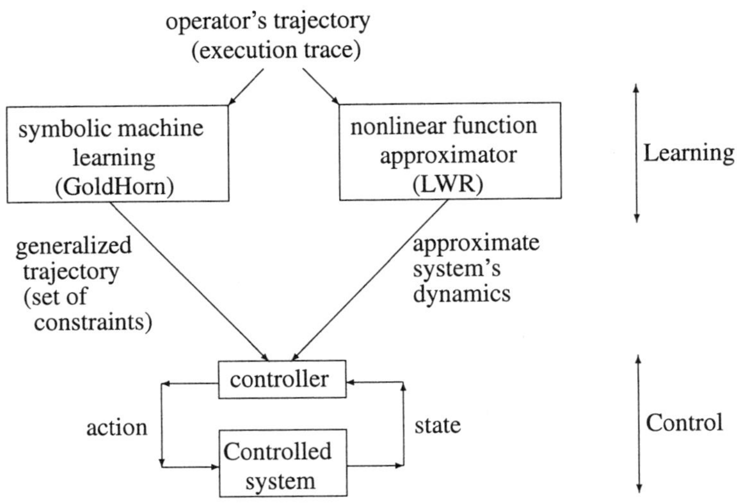


Figure 4.1: Reconstructing human control skill from operator's trajectory.

in Section 4.1.2.

Experiments later in this chapter show that such a qualitative strategy is easy to understand and can be used to explain the operator's subconscious control skill. Moreover, it defines a family of quantitative strategies and provides a *space for controller optimization.* By transforming a qualitative strategy into a set of operational quantitative strategies we experimentally show that qualitative strategies can be general (not sensitive to the exact settings of the qualitative parameters) and very successful. The methodology used for such a transformation is given in Section 4.1.3.

### 4.1.1 Symbolic Skill Reconstruction using GoldHorn

Learning the operator's control skill is decomposed (see Figure 4.1) into generalizing the operator's trajectory by equation learning program GoldHorn and learning the system's dynamics by locally weighted linear regression. Both, the generalized trajectory and the nonlinear model of system dynamics are then used to control the system. The use of the domain knowledge with this approach is illustrated in the experiments and discussed in depth in Chapter 7.

The generalized operator's trajectory is induced by GoldHorn from one or more execution traces. We do not learn the trajectory in time, but rather induce constraints among the state variables in the operator's execution trace (or traces). Control actions are deliberately left out from the traces, since they can be computed if we know the desired next state and have an approximate model of the system. Similarly, we can leave out some of the state variables, (typically the time derivatives of other state variables), to keep the induced constraints as simple as possible. Of course, this requires some domain knowledge. GoldHorn then induces constraints among the remaining variables in the form of ordinary differential equations and ranks them according to their significance judged by two measures: the multiple correlation coefficient and the normalized deviation. One or more of the most significant induced equations are used as constraints which define the generalized operator's trajectory. Note, that there are cases when a single equation does not uniquely define the generalized trajectory. In such a case it is possible to consider other equations as all of them reflect the constraints between the state variables along the operator's trajectory. One such example of combining equations is presented in the experiments with the acrobot in Section 4.3.

The generalized trajectory and the learned model of the system dynamics form an indirect controller that can be used to control the system as formally described in Section 3.1. Ideally, to mimic the operator's control strategy, the constraints should hold in every state. So an appropriate control action should minimize the deviation of the predicted next state from the constraints. A measure of such deviation of a state from the constraints will be called *constraint error*. A constraint error simply measures the degree to which a system's state does not satisfy the constraints. Given a system state and the constraints, an appropriate action can be computed in many ways. In general it requires a model of the system's dynamics and minimization of the constraint error over the possible actions (as defined in Equation 3.2). If it is possible to express one state variable as a function of other state variables in the constraints, the minimization of error in this chosen (dependent) state variable can be used to determine the action to take. We chose a dependent variable that is, according to the human notion of causality in the system, directly affected by the available control actions. The selection of the trajectory variables is discussed in depth in Chapter 7.

The minimization of constraint error over all actions might be computationally expensive when the set of possible actions is large. The use of LWR enables that an appropriate action is computed directly as pointed out in Section 3.1.2. At each current state we learn locally linear model expressing dependent trajectory variables in the next state given the action and the current state. Then we use the inverse of the linear model to compute the action aiming to achieve the desired value of dependent variables. So, the actual controller consists of two parts: nonlinear model of the system's dynamics (which is updated online) and a desired trajectory. At each time step the controller considers the current state, estimates a local linear model around the current state and computes an action which makes one step towards the desired next state on the trajectory.

Note that a potential problem, of generalizing the operator's trajectory as the constraints among the state variables in a single state, is that following the generalized trajectory as a continuously changing subgoal could cause the system to move backwards in time. However, moving backwards in time by minimizing the error with respect to trajectory constraints does not happen if the system's dynamics do not permit the previous state to be reached by a single action. The acrobot is an example of such a system. For domains in which this would be a problem, the alternative approach is to generalize the constraints between the current and the next state in the execution trace. This second approach was used in the crane domain.

### 4.1.2 From Equations to Qualitative Strategy

The generalized trajectory in the form of equations induced by GoldHorn defines *quantitative strategy* and can be used to control the system. Also, it provides some insight into the operator's control strategy. Since equations are often hard to interpret, we use qualitative abstraction to transform the induced quantitative strategy into a *qualitative strategy*. This qualitative strategy is expressed using *qualitative terms* and *qualitatively constrained functions*. A qualitative term is an abstraction of a numerical value or a numerical constraint. Qualitatively constrained functions are monotonicity constraints as used in qualitative reasoning with variants used also in Qualitative Process Theory (QPT, Forbus, 1984) and qualitative simulation (QSIM, Kuipers, 1986). Qualitatively constrained functions are more thoroughly discussed in Chapter 5 (page 70). Here we give just the necessary definitions.

A *qualitatively constrained function* (abbreviated QCF) $M^{s_1,\ldots,s_m} : \Re^m \mapsto \Re$, $s_i \in \{+,-\}$ represents an arbitrary function with $m$ continuous arguments that respect the qualitative constraints given by signs $s_i$. The qualitative constraint given by sign $s_i = +$ ($s_i = -$) requires that the function is strictly increasing (decreasing) in its dependence on the $i$-th argument. We say that the function is *positively related* (negatively related) to the $i$-th argument. $M^{s_1,\ldots,s_m}$ represents any function which is, for all $i = 1,\ldots,m$ positively (negatively) related to the $i$-th argument, if $s_i = +$ ($s_i = -$). In the case of continuously derivable functions, $s_i = +$ ($s_i = -$) requires that the $i$-th partial derivative is strictly positive (negative):

$$
\begin{aligned}
&M^{s_1,\ldots,s_m} : \Re^m \mapsto \Re\,,\ s_i \in \{+,-\} \\
&\frac{\partial M^{\ldots,s_i,\ldots}(\ldots,x_i,\ldots)}{\partial x_i} \begin{cases} > 0, \text{if } s_i = + \\ < 0, \text{if } s_i = - \end{cases}
\end{aligned}
\tag{4.1}
$$

Note, that the qualitative constraint given by sign $s_i = +$ only states that when the $i$-th argument increases, the QCF will also increase, *barring other changes*. It can happen that a QCF with the constraint $s_i = +$ decreases even if the $i$-th argument increases, because of a change in another argument. For example, consider the behaviour of gas pressure in a container: $Pres = c \times Temp/Vol$, where $c$ is a positive constant. We can express the qualitative behaviour of gas pressure by QCF $Pres = M^{+,-}(Temp, Vol)$. This constraint allows that the pressure decreases even if the temperature increases, because of the change in the volume. Note, however, that the qualitative behaviour of gas is not consistent with the constraint $Pres = M^{+}(Temp)$.

According to the definition of QCFs, a continuous function $f$, can be abstracted to a QCF by computing the partial derivatives. Positive (negative) $\frac{\partial f}{\partial x_i}$ corresponds to a QCF that is positively (negatively) related to the $x_i$. In the case of gas pressure, QCF $Pres = M^{+,-}(Temp, Vol)$ is derived from equation $Pres = c \times Temp/Vol$, $c > 0$ as follows:

$$
\begin{aligned}
\frac{\partial Pres}{\partial Temp} &= \frac{c}{Vol} > 0 \\
\frac{\partial Pres}{\partial Vol} &= \frac{-c \times Temp}{Vol^2} < 0
\end{aligned}
\tag{4.2}
$$

### 4.1.3 Transforming Qualitative Strategy into Quantitative Strategies

A qualitative strategy defines a family of quantitative strategies. One motivation to transform a qualitative strategy into quantitative strategies is to verify if the qualitative strategy gives any useful control knowledge and reflects the operator's control strategy. The second, more interesting motivation is to use a qualitative strategy as a space for controller optimization. It is important to note that we can use additional constraints imposed by the domain knowledge and in this way reduce the optimization space. This will be illustrated later, in experiments in the crane and acrobot domains.

A qualitative strategy is expressed in terms of qualitative terms and qualitatively constrained functions. To use it to control the system, qualitative terms have to be concretised into numerical values or constraints and QCFs have to be transformed into real-valued functions respecting the qualitative constraints. In the next paragraphs we describe how piece-wise linear QCFs can be generated at random.

We define a set of functions $F^{s_1,\ldots,s_m}$, $s_i \in \{+,-\}$ defined on $[0,1]^m$ and mapping to $[0,1]$, such that any function $f^{s_1,\ldots,s_m}$ from this set is monotonically increasing (decreasing) in the $i$-th argument, if $s_i = +$ ($s_i = -$). Different piece-wise linear functions $f^{s_1,\ldots,s_m}$ can be generated by splitting the space $[0,1]^m$ into $C = \prod_{i=1}^{m} R_i$ ($R_i$ is the number of slices in the $i$-th dimension) cells and assigning each cell (with address $a = (a_1,\ldots,a_m)$, $a_i \in \{0,\ldots,R_i - 1\}$) an appropriate value $v_a$ in its left-bottom corner. The values at $x_i = 1$ are addressed by $a_i = R_i$. The values at other points in a cell are linearly interpolated by considering the values at $m$ neighbours of the cell. The value at a point $x$, with the cell address $a = (a_1\ldots,a_m)$, $a_i = \lfloor x_i R_i \rfloor$ can be computed as:

$$f^{s_1,\ldots,s_m}(x) = v_a + \sum_{i=1}^{m} d_i(a)(x_i R_i - \lfloor x_i R_i \rfloor)$$

Here $d_i(a) = v_{\ldots,1+a_i,\ldots} - v_{\ldots,a_i,\ldots}$ denotes the $i$-th neighbour difference. To satisfy the qualitative constraints, $sign(d_i(a)) = s_i$.

Suppose we want to generate such random functions $f^{s_1,s_2}$ on a $2 \times 2$ grid, and the values of the edge cells are already stored as $v_{0,0}$, $v_{0,2}$, $v_{2,0}$ and $v_{2,2}$. We first set the values $v_{1,0}$ and $v_{1,2}$ by taking two "suitable" random values from $(v_{0,0}\ldots v_{2,0})$ and $(v_{0,2}\ldots v_{2,2})$. These two values are suitable if their differences respect the constraints given by the signs $s_1, s_2$. Similarly we set the values $v_{0,1}$ and $v_{2,1}$. The value of the middle cell $v_{1,1}$ is set to a random number from $(v_{1,0}\ldots v_{1,2}) \cap (v_{0,1}\ldots v_{2,1})$. Note that the intersection is not empty, if the edge values are suitable. Similarly, random functions $f^{s_1,\ldots,s_m}$ can be generated by dividing the space into $m^2$ subregions, generating suitable values for the edge cells in every region, and recursively set the cell values in every subregion.

Note that here defined piece-wise linear QCFs are defined on $[0,1]^m$ and mapping to $[0,1]$. It is easy to adopt the range and the definition area as needed. For example, piece-wise linear QCF $g^+$ defined on $[0,p]$ and mapping to $[0,r]$ is obtained as $g^+(x) = r \times f^+(x/p)$.

## 4.2 Skill Reconstruction in Crane Domain

The reconstruction of human crane control skill generally follows the approach described in Section 4.1. Here we give details specific to the crane domain. This domain is described in Section 2.3.1.

In experiments with humans learning to control the crane simulator, remarkable individual differences were observed in the speed of control and the characteristics of the strategies they used. Some operators tended towards fast and less reliable operation, others were more conservative and slower, in order to avoid large rope oscillations. Ideally, we would be able to reconstruct these individual differences between the operators in the style of driving the crane. In previous experiments with behavioural cloning using regression trees (Urbančič and Bratko, 1994), the main problem for trees was the swing control, i.e. large rope oscillations when the trolley approached its goal position. For this reason slower and more conservative traces were found by far the most useful for cloning with regression trees. On the other hand, strategies of faster operators are more complex, requiring the exact timing and skill to decrease large rope oscillations due to large trolley accelerations. Here we use traces of the fastest operator, that is the operator who achieved shortest times. He was able to afford large initial swing caused by large acceleration, and later skillfully reduce the swing.

First we give the details of learning operator's quantitative strategy and give an example of the induced strategy. This quantitative strategy is then abstracted to a qualitative strategy that provides a good insight into the operator's control skill. By transforming the qualitative strategy into operational quantitative strategies in Section 4.2.3 we show that the qualitative strategy is general, robust and can be used to optimize operator's control strategy. We also demonstrate how the domain knowledge can be combined with a qualitative strategy.

### 4.2.1 Learning Quantitative Strategy

We used GoldHorn to induce the generalized operator's trajectory from one or more traces of the same subject. In the case of the crane control, two forces are applied to the system: the force to the trolley $F_X$ and the force to the rope $F_L$. We choose to define the generalized by two variables: the desired trolley velocity $\dot{X}_{des}$ and the desired rope length $L_{des}$. Of course, generalized trajectory could be expressed in a different way, for example ($\dot{X}_{des}$,$\dot{L}_{des}$). We here decided for ($\dot{X}_{des}$,$L_{des}$) trajectory. The choice of this variables is discusses in depth in Chapter 7. The learned trajectory is given by $\dot{X}_{des}$ and $L_{des}$ as a function of other state variables, that is $\dot{X}$ and $L$ in the next state as the function of the current state. In fact, $\dot{X}_{des}$ and $L_{des}$ are learned separately, but they form one single trajectory since they are functions of the same variables. It is important to note that we do not learn the ($\dot{X}_{des}$,$L_{des}$) trajectory in time, but rather constraints between the state variables in the form of differential equations induced by GoldHorn from the execution trace. GoldHorn uses the multiple correlation coefficient and the normalized deviation to judge the significance of an equation. Most significant equations are used to express the desired $\dot{X}_{des}$ and $L_{des}$ as a function of the current state.

When the trajectory is known, the action which achieves the desired next state on the trajectory is computed using a locally linear model learned by LWR from some execution traces, as described in Section 4.1.1. At each time step the controller considers the current state and computes the desired trolley velocity $\dot{X}_{des}$ and the desired rope velocity $L_{des}$ according to the induced equations defining the generalized trajectory. Afterwards it learns a model of system dynamics near the current state by LWR and uses this approximate model to select actions $F_X$ and $F_L$ aiming to achieve the desired $\dot{X}_{des}$ and $L_{des}$. Precisely, LWR learns local linear model near the current state, i.e. $\dot{X}$ and $L$ at next times step as a linear function of the state variables and the corresponding force:

$$\dot{X}(t+dt) = k_1X(t) + k_2\dot{X}(t) + k_3\Phi(t) + k_4\dot{\Phi}(t) + k_5L(t) + k_6\dot{L}(t) + k_7F_X(t)$$
$$L(t+dt) = l_1X(t) + l_2\dot{X}(t) + l_3\Phi(t) + l_4\dot{\Phi}(t) + l_5L(t) + l_6\dot{L}(t) + l_7F_L(t)$$

Actions $F_X$ and $F_L$, aiming to achieve the desired $\dot{X}_{des}(t)$ and $L_{des}(t)$ are then computed from the linear model:

$$F_X = (\dot{X}_{des} - k_1X(t) - k_2\dot{X}(t) - k_3\Phi(t) - k_4\dot{\Phi}(t) - k_5L(t) - k_6\dot{L}(t))/k_7$$
$$F_L = (L_{des} - l_1X(t) - l_2\dot{X}(t) - l_3\Phi(t) - l_4\dot{\Phi}(t) - l_5L(t) - l_6\dot{L}(t))/l_7$$

This approach to skill reconstruction turned out to be generally very effective. For every successful execution trace we experimented with, GoldHorn induced a generalized trajectory which allowed the controller to complete the task. The induced control strategy often

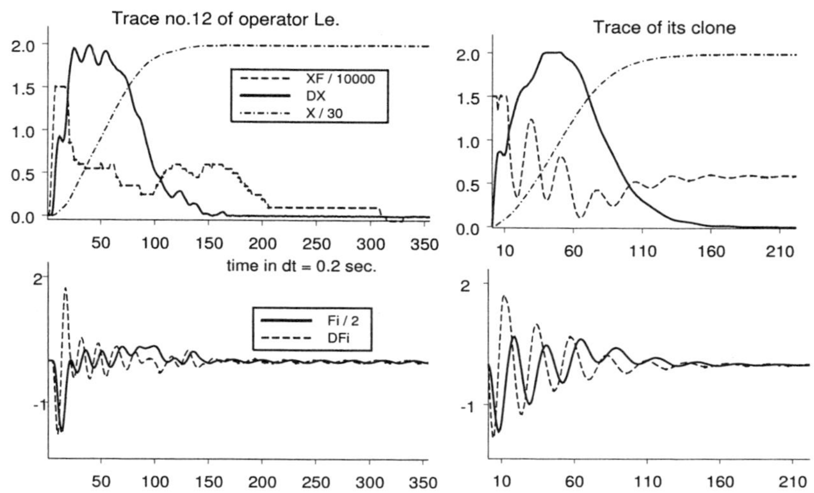


Figure 4.2: Original and its clone: on the left is the original trace and on the right is the trace of its clone, using the induced quantitative strategy given in Equations 4.4 and 4.5. The original concludes the task in 72 sec. The clone is much faster and concludes the task in 44 sec. Decreasing of the rope swing can be observed in both traces.

concludes the task faster than the original operator's traces. However, the time needed to complete the task varied significantly with example execution trace and the particular differential equation used. We found out that differential equations with linear and quadratic factors were usually the most successful. Terms of higher orders usually did not improve the performance. Another interesting, but not surprising observation was, that usually a more successful trajectory was induced from a faster execution trace. So, individual differences between the operators, like differences in the operator's speed and skill to decrease the swing of the rope, were reflected in the generalized trajectories induced from their traces.

### An Example of Learned Quantitative Strategy

In the next paragraphs we present the strategy learned from one of the traces of the fastest operator (see Figure 4.2). To generalize the operator's $(\dot{X}_{des}, L_{des})$ trajectory GoldHorn was used to learn the constraints in the trace among the state variables as differential equations with linear or quadratic terms.

The desired trajectory of the trolley velocity $\dot{X}_{des}$ (Equation 4.4) was expressed from the most significant equation (Equation 4.3), induced from the operator's trace:

$$509.52 + 51.24X + 28.21\Phi - 564.32\dot{X}_{des} - X^2 = 0 \tag{4.3}$$

$$\dot{X}_{des} = 0.902 - 0.0018X^2 + 0.090X + 0.050\Phi \tag{4.4}$$

To control the rope, we use a very simple equation with higher error estimate. As discussed also in Section 7.2.2, the most significant equation successfully controls the rope, but here we use a very simple equation that was also induced by GoldHorn to show a possible

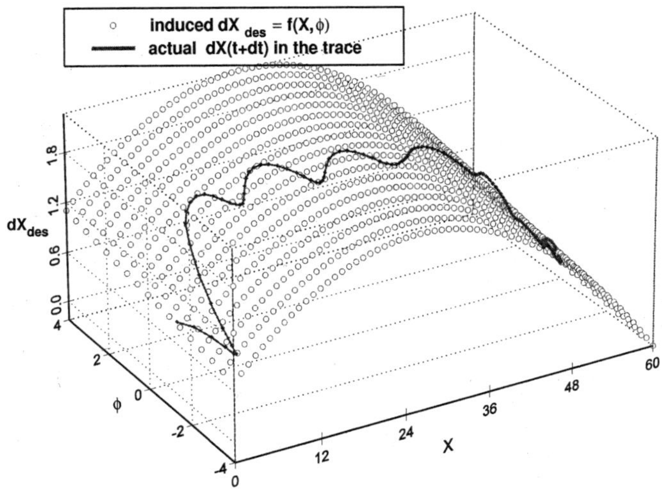


Figure 4.3: Induced quantitative strategy for the trolley control: the dotted surface represents the induced trajectory $\dot{X}_{des}$ (Equation 4.4) as a function of $X$ and $\Phi$. The solid curve represents the actual $\dot{X}(t + dt)$ from the original operator's trace. Note that $\dot{X}_{des}$ is first (when $X < 25.62$) positively related to $X$ and later negatively related to $X$. Also $\dot{X}_{des}$ is positively related to $\Phi$, revealing operator's skill to decrease the swing of the rope.

use of the domain knowledge in equation analysis. This simple equation with higher error estimate is:

$$L_{des} = 0.0037X^2 + 17.46$$

This equation is simple, but not accurate enough, since it states that $L_{des}$ at the goal position $X_g$ is 30.7. To conclude the task, the rope length at the goal position should be $L_{des}(X_g)=L_g$=32. This can be corrected by adding this difference to the learned $L_{des}$. In this way we get the rule to control the rope length:

$$L_{des} = 0.0037X^2 + 17.46 + 1.3 \tag{4.5}$$

Although this rule is simple, it is very successful, robust and easy to understand: the operator brings down the load as the trolley moves from its start to the goal position.

The rules given by Equations 4.4 and 4.5 describe the generalized operator's trajectory. The actual controller consists of the generalized trajectory and the learned system's dynamics. The controller was tested on the simulator. It concludes the task in 44 sec. and is much faster than the original operator's trace, which needs 72 sec. A crucial part of the crane control strategy, that is decreasing of the rope swing, can be observed in both traces (see Figure 4.2). This is done by the acceleration of the trolley when the rope angle increases, and can also be observed in the induced $\dot{X}_{des}$ in Figure 4.3.

### Notes on Experiments

We used traces sampled at sample time $dt = 0.2$ sec. $\dot{X}_{des}$ and $L_{des}$ were computed as $\dot{X}_{des}(t) = \dot{X}(t + dt)$ and $L_{des}(t) = L(t + dt)$. To learn the desired trajectory of the trolley velocity $\dot{X}_{des}$, GoldHorn was used to learn the constraints in the trace among variables $\dot{X}_{des}$, $X$, $\Phi$ and $\dot{\Phi}$ as differential equations with linear or quadratic terms. The trajectory of rope length $L_{des}$ was learned similarly by using variables $L_{des}$, $X$, $\dot{X}$, $\Phi$ and $\dot{\Phi}$. The choice of the trajectory variables is discussed in depth in Chapter 7. To learn the system's dynamics with LWR the original trace and another two traces of the same operator were used.

The traces were preprocessed: the first second of the trace (5 learning examples for GoldHorn) was excluded from the trace, since the trace usually starts with a few zero actions ($F_X = F_L = 0$), due to the operator's delay. These zero actions are followed by actions ($F_X = 500$,$F_X = 1000$, $F_X = 1500$,..), where operator tries to increase the horizontal force in order to move the trolley. This takes a few steps, due to the operator's limitation to increase the horizontal force in steps of 500. Preprocessing the traces is justified by the observation that, including first "zero–action" examples in the learning examples is just adding noise, since these examples imply that at $X = 0$ (starting position of the trolley) operator does not use any action ($\dot{X}_{des} = 0$) – this is only the result of the operator's delay. Moreover, the zero–action states and the first nonzero–action state are exactly the same, but they imply different actions.

### 4.2.2 Qualitative Strategy

The induced quantitative strategy can be generalized into a qualitative strategy as described in Section 4.1.2. Since controlling the trolley is considerably harder than controlling the rope, we focus on $\dot{X}_{des}$ trajectory. Of course, the same approach as with $\dot{X}_{des}$ trajectory can be used to generalize $L_{des}$ trajectory into a qualitative strategy for the rope control.

Qualitative constraints defining a qualitative strategy are derived by the qualitative abstraction of the quantitative strategy given by Equation 4.4:

$$\begin{aligned}
\frac{\partial \dot{X}_{des}}{\partial \Phi} &= 0.050 > 0 \\
\frac{\partial \dot{X}_{des}}{\partial X} &= \frac{-2X + 51.24}{564.32} \\
&\Rightarrow \dot{X}_{des} = M^{+,+}(X, \Phi) \text{ if } X < 25.62 \\
&\quad\ \dot{X}_{des} = M^{-,+}(X, \Phi) \text{ if } X > 25.62
\end{aligned}$$

giving the qualitative strategy for the trolley control:

$$\begin{aligned}
&\text{if } X < X_{mid} \text{ then} \\
&\quad \dot{X}_{des} = M^{+,+}(X, \Phi) \\
&\text{else } \dot{X}_{des} = M^{-,+}(X, \Phi)
\end{aligned} \tag{4.6}$$

Note that $X_{mid}$, $M^{+,+}(X, \Phi)$ and $M^{-,+}(X, \Phi)$ are qualitative parameters of the strategy. $M^{+,+}(X, \Phi)$ and $M^{-,+}(X, \Phi)$ are qualitatively constrained functions (abbreviated QCFs). Qualitative term $X_{mid}$ is a qualitative abstraction of the value 25.62.

This qualitative strategy provides a good insight into the operator's control strategy. First, when the trolley is near the start position ($X < X_{mid}$), $\dot{X}_{des}$ is positively related to $X$, since the operator increases the trolley velocity. Afterwards the operator decreases the trolley velocity to approach the goal. The most interesting is a qualitative constraint stating that the trolley velocity is positively related to the rope angle. This is a very important part of the operator's strategy and describes his/her skill to decrease the swing of the rope. By the acceleration of the trolley when the rope angle increases the angular velocity is decreased.

The qualitative strategy given by Rule 4.6 defines a set of quantitative strategies for the trolley control. One instance is also the operator's quantitative strategy from Equation 4.4, with the qualitative parameters:

$$
\begin{aligned}
&X_{mid} = 25.62 \\
&M^{+,+}(X,\Phi) = 0.902 - 0.0018X^2 + 0.090X + 0.050\Phi, \text{if } X < 25.62 \\
&M^{-,+}(X,\Phi) = 0.902 - 0.0018X^2 + 0.090X + 0.050\Phi, \text{if } X > 25.62
\end{aligned}
$$

### 4.2.3 Transforming Qualitative Strategy into Quantitative Strategies

The qualitative strategy, given by Rule 4.6, defines a set of quantitative strategies $S$ with free qualitative parameters $X_{mid}$, $M^{+,+}(X,\Phi)$ and $M^{-,+}(X,\Phi)$. To transform the qualitative strategy into an operational quantitative strategy, the qualitative parameters have to be concretised into quantitative values or functions. Clearly, some of these strategies are not successful. For example, if $M^{+,+}(0,\Phi) < 0$ the trolley moves away from the goal. Here we investigate how successful and general the learned qualitative strategy is, that is how sensitive it is to changes of the qualitative parameters. In order to do this we performed two sets of experiments:

- In the first experiment we used just basic knowledge of the control task (like goal position and allowed ranges of the state variables) and randomly generated functions satisfying the qualitative constraints. Results show that the qualitative strategy is general and that the exact selection of the qualitative parameters is not crucial for its success.
- In the second experiment we restricted QCFs to simple (quadratic) functions, and used knowledge of the control task to find constraints among parameters of the quantitative strategies. We experimentally confirm that by using some knowledge of the control task it is easy to find the constraints among parameters of the quantitative strategies which filter out bad strategies. Very successful strategies can be obtained in this way. Since the generalized trajectory is expressed in a symbolic form, it is easy to understand, analyze and correct.

Since controlling the trolley is considerably harder than controlling the rope, we focus on the strategy for the trolley control. Of course, the same approach as with $\dot{X}_{des}$ trajectory could be used to generalize $L_{des}$ trajectory into a qualitative strategy and to transform it into a set of quantitative strategies for the rope control. In all experiments we used $L_{des}$ trajectory given by Equation 4.5.

#### Using Basic Knowledge and Randomly Generated QCFs

Here we use basic knowledge of the control task and randomly generated functions with the required qualitative properties, to transform the qualitative strategy into operational quantitative strategies. The knowledge used consists of the task limits ($X < X_{max}$, $|\Phi| < \Phi_{max}$, $\left|\dot{X}_{des}\right| < \dot{X}_{max}$) and two simple facts of the control task:

1. The trolley starts towards the goal: $0 < \dot{X}_{des}(0,0) \leq \dot{X}_{max}$.

2. The trolley stops at $X_g$: $\dot{X}_{des}(X_g, 0)$=0.

First we show how this domain knowledge can be combined with the qualitative strategy from Rule 4.6 to define a set of quantitative strategies. Then experimental results are presented.

For simplicity of combining the domain knowledge and the qualitative strategy we replace each of the QCFs as the sum of two QCFs:

$$M^{+,+}(X, \Phi) = M^+_X(X) + M^+_{\Phi 1}(\Phi)$$
$$M^{-,+}(X, \Phi) = M^-_X(X) + M^+_{\Phi 2}(\Phi)$$

By doing this replacement, we lose some generality of the qualitative strategy, but we preserve the correspondence with the induced quantitative strategy given by Equation 4.4. Let $f^+$ denote strictly increasing real-valued functions defined on $[0, 1]$ and mapping to $[0, 1]$, and let $h^+$ ($h^-$) denote strictly increasing (decreasing) functions defined on $[-1, 1]$ and mapping to $[-1, 1]$, with the property that $h^+(0) = h^-(0) = 0$. Since $|\Phi| < \Phi_{max}$, and since we can (without loss of generality) assume that one QCF in the sum of QCFs is crossing zero, increasing functions $M^+_{\Phi 1}$ and $M^+_{\Phi 2}$ can be written as:

$$M^+_{\Phi 1}(\Phi) = k_1 h^+_1(\frac{\Phi}{\Phi_{max}})$$
$$M^+_{\Phi 2}(\Phi) = k_1 h^+_2(\frac{\Phi}{\Phi_{max}})$$

Since $\left|\dot{X}_{des}\right| < \dot{X}_{max}$, $k_1 \in (0, \dot{X}_{max})$. Because the trolley must start towards goal ($\dot{X}_{des}(0,0) > 0$) and $\left|\dot{X}_{des}\right| < \dot{X}_{max}$, $M^+_x$ can be written as:

$$M^+_x(X) = \dot{X}_{max} f^+_1(X/X_{mid})$$

Since the trolley must stop at the goal position ($\dot{X}_{des}(X_g, 0)$=0), $M^-_x(X)$ can be written as:

$$M^-_x(X) = \dot{X}_{max} h^-_3(\frac{X - X_g}{X_g - X_{mid}})$$

This derivation gives a set of quantitative strategies for the trolley control $S_{q0} \subset S$:

$$\begin{aligned} &\text{if } X < X_{mid} \text{ then} \\ &\quad \dot{X}_{des} = \dot{X}_{max} f^+_1(\frac{X}{X_{mid}}) + k_1 h^+_1(\frac{\Phi}{\Phi_{max}}) \\ &\text{else } \dot{X}_{des} = \dot{X}_{max} h^-_3(\frac{X - X_g}{X_g - X_{mid}}) + k_1 h^+_2(\frac{\Phi}{\Phi_{max}}) \\ &X_{mid} \in (0, X_g),\ k_1 \in (0, \dot{X}_{max}) \end{aligned} \tag{4.7}$$

These quantitative strategies were tested on the simulator. $X_{mid}$ was set to values 15, 30, 45 and $k_1$ to values $\dot{X}_{max}/2$, $\dot{X}_{max}/4$ and $\dot{X}_{max}/8$. For each of these parameter settings, we randomly generated 10 sets of piece-wise linear functions $f^+_1$, $h^+_1$, $h^+_2$, $h^-_3$. Functions $f^+$ were generated by taking $R + 1$ random numbers sorted in increasing order $0 \le y_i \le 1$, $i = 0, \cdots, R$ ($y_i < y_{i+1}$) and set $f^+(i/R)$ to $y_i$. The values at other points were linearly

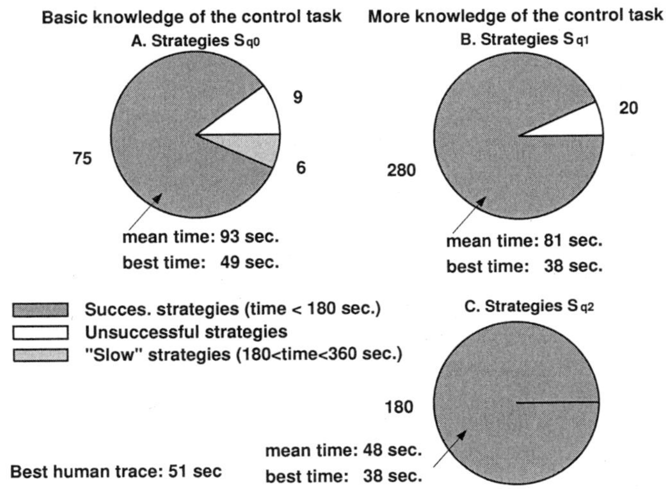


Figure 4.4: Efficiency of qualitative strategy for the trolley control: Pie A. gives results with randomly generated QCFs. Out of 90 different quantitative strategies from $S_{q0}$, 75 finish the task in 180 sec. or less. Another 6 are successful in 360 sec. Pies B. and C. give results with using more domain knowledge. All of the 180 strategies from $S_{q2}$ are successful and very fast.

interpolated: $f^+(x) = y_j + (y_{j+1} - y_j)(Rx - j)$ where $j = \lfloor xR \rfloor$. Functions $h^+$ ($h^-$) were generated similarly by taking $R/2$ random numbers from $[-1, 0)$, number 0, and $R/2$ random numbers from $(0, 1]$ and sort them in increasing (decreasing) order. All piece-wise linear functions $f_1^+$, $h_1^+$, $h_2^+$, $h_3^-$ were generated using $R = 10$. Note that this procedure is slightly different than the procedure described in Section 4.1.3. We expect similar results, if there described procedure to generate piece-wise linear QCFs at random would be used.

In this way we get a set of $3 \times 3 \times 10 = 90$ different quantitative strategies from set $S_{q0}$. Considering the fact that we used a qualitative strategy, where QCFs were generated at random and we used only the basic knowledge of the crane task, results are very good. Out of 90 different quantitative strategies 75 finished the task in 180 sec. or less. 39 strategies completed the task in less than 80 sec., some of them faster than 50 seconds. The details are given on Figure 4.4.

### Using More Domain Knowledge

The qualitative strategy (Rule 4.6) defines a set of quantitative strategies $S$. Here we restrict qualitative constrained functions to quadratic functions and use the knowledge of the control task to find constraints among parameters of the quantitative strategies $S$. In order to do this, we define quantitative strategies $S_{q1}$ as a subset of strategies $S$ with the properties:

1. *QCFs $M^{+,+}(X, \Phi)$ and $M^{-,+}(X, \Phi)$ with quadratic dependence on $X$ and linear dependence on $\Phi$.* Dependence on $\Phi$ is restricted to linear functions $k\Phi + n$, and dependence on $X$ is restricted to quadratic functions $pX^2 + rX + m$. We use quadratic functions as a tradeoff between simplicity and variety in the resulting strategies. The results show that also simpler linear functions (obtained when $p_{1,2} = 0$) are successful.

Parameters $n$ and $m$ are (without loss of generality) joined into $\dot{X}_{mid}$. This defines a subset of quantitative strategies of the form:

$$\begin{aligned}
&\text{if } X < X_{mid} \text{ then} \\
&\quad \dot{X}_{des} = \dot{X}_{mid1} + p_1 X^2 + r_1 X + k_1 \Phi \\
&\text{else } \dot{X}_{des} = \dot{X}_{mid2} + p_2 X^2 + r_2 X + k_2 \Phi
\end{aligned} \tag{4.8}$$

2. *Satisfy the task requirements* (obey task limits, stay at the goal position).

3. *Start towards goal* ($\dot{X}_{des}(0,0) > 0$).

4. *Continuous and fast at* $X_{mid}$ ($\dot{X}_{des}(X_{mid} - \epsilon, 0)$=$\dot{X}_{des}(X_{mid} + \epsilon, 0)$ =$\dot{X}_{max}$).

Each of these properties imposes constraints on the parameters. The derivation of the constraints on the parameters given in Section 4.2.4 yields in the strategies $S_{q1}$ with the form given by Rule 4.8 and constraints:

$$\begin{aligned}
&0 < X_{mid} < X_g \,,\, 0 < \dot{X}_{mid1} < \dot{X}_{max} \\
&-\frac{\dot{X}_{max}}{\Phi_{max}} < k_1 = k_2 < \frac{\dot{X}_{max}}{\Phi_{max}} \\
&-\frac{\dot{X}_{max} - \dot{X}_{mid1}}{X_{mid}{}^2} < p_1 < \frac{\dot{X}_{max} - \dot{X}_{mid1}}{X_{mid}{}^2} \\
&-\frac{\dot{X}_{max}}{(X_g - X_{mid})^2} < p_2 < \frac{\dot{X}_{max}}{(X_g - X_{mid})^2} \\
&r_1 = \frac{\dot{X}_{max} - \dot{X}_{mid1}}{X_{mid}} - p_1 X_{mid} \,;\; r_2 = \frac{\dot{X}_{max}}{X_{mid} - X_g} - p_2(X_{mid} + X_g) \\
&\dot{X}_{mid2} = -X_g(X_g p_2 + r_2)
\end{aligned} \tag{4.9}$$

Further we define strategies $S_{q2}$ as a subset of strategies $S_{q1}$ with additional property that *the trolley travels fast*. This constraint simply requires that the velocity of the trolley is large enough, when the trolley is far from the goal position. Constraints on parameters for strategies $S_{q2}$ are the same as for $S_{q1}$ with additional constraint on $p_2$, derived in Section 4.2.4.

Note that the last two constraints (*continuous and fast at* $X_{mid}$ and *the trolley travels fast*) are not necessary. If we do not use them, strategies are still successful in terms of reaching the goal position under the task limits, but some of them are too slow and fail to conclude the task in the required time of 180 seconds.

To investigate how successful those strategies are, we tested strategies $S_{q1}$ with different values of parameters: 2 for $k_1$ and $\dot{X}_{mid1}$, 3 for $X_{mid}$ and 5 for $p_1$ and $p_2$. Parameters were taken uniformly spaced (except for $p_1$ and $p_2$) from the parameter's range given by Equation

4.9:

$$
\begin{aligned}
k_1 &= \frac{\dot{X}_{max}}{4\Phi_{max}}, \frac{3\dot{X}_{max}}{4\Phi_{max}} \\
\dot{X}_{mid1} &= \frac{\dot{X}_{max}}{3}, \frac{2\dot{X}_{max}}{3}; \quad X_{mid} = \frac{X_{max}}{4}, \frac{2X_{max}}{4}, \frac{3X_{max}}{4} \\
p_1 &= \frac{m_1}{5} \frac{\dot{X}_{max} - \dot{X}_{mid1}}{X_{mid}^{\ 2}}, m_1 = -5, -3, 0, 3, 5 \\
p_2 &= \frac{m_2}{5} \frac{\dot{X}_{max}}{(X_g - X_{mid})^2}, m_2 = -5, -3, 0, 3, 5
\end{aligned} \tag{4.10}
$$

These parameter settings give 300 different qualitative strategies from $S_{q1}$. 180 of these strategies obey additional constraint *the trolley travels fast* and belong to $S_{q2}$. All of these 180 different quantitative strategies are successful and very fast. The best time is 38 sec., mean time 48 sec. and the worst time 78 sec. Considering the fact that the fastest operator's trace completes the task in 51 sec., the results are excellent. Out of other 120 strategies, which do not respect the constraint *the trolley travels fast* and do not belong to $S_{q2}$, 100 are successful with mean time 84 sec. The details are given in Figure 4.4. Also the most simple strategies from $S_{q1}$ with linear dependence in $X$ and $\Phi$ that were obtained with parameters $p_{1,2} = m_{1,2} = 0$ were successful. All of the 12 corresponding strategies tested were successful with average time 62 sec.

### Test of Robustness

The strategies $S_{q0}$, were also tested for their robustness against changes in the start state, the test where the controllers in the form of decision or regression trees usually fail. In these experiments $X_{mid}$ was set to values 15, 30, 45 and $k_1$ to $\dot{X}_{max}/4$. For each of these parameters settings, we generated 10 sets of random functions $f_1^+$, $h_1^+$, $h_2^+$, $h_3^-$ with $R = 10$. In this way we get 30 quantitative strategies. All of these strategies were tested for their robustness against changes in the start state. Starting horizontal position was modified by 10% and 50%, and starting rope angle by -20%, 0% and 20%. This amounts to six combinations – six different tasks. Since controllers based on generalized operator's trajectory are goal directed, the change of the start state didn't affect their performance significantly. Out of $30 \times 6 = 180$ combinations, 162 were successful. The percentage of successful strategies with the changed start state is the same as the percentage of successful strategies with the original start state.

These results show significant improvement in the robustness of controllers using the generalized operator's trajectory over the controllers in the form of decision or regression trees.

## 4.2.4 Derivation of Constraints on Qualitative Parameters

In Section 4.2.3 we used knowledge of the crane control task to transform a qualitative strategy into operational quantitative strategies. Here we give the derivation of constraints on the qualitative parameters $X_{mid}$, $M_x^+$, $M_x^-$, $M_{\Phi 1}^+$ and $M_{\Phi 2}^+$ for quantitative strategies $S_{q1}$ and $S_{q2}$, which have the general form given by Rule 4.8. Each property of these strategies gives constraints on qualitative parameters:

1. *Quadratic $M_x^+$ and $M_x^-$ and linear $M_{\Phi 1}^+$ and $M_{\Phi 2}^+$.* Since functions have to be increasing or decreasing we use constraints on their derivatives:

$$\underline{k_1, k_2 > 0}$$
$$2p_1 X + r_1 > 0 \text{ if } X < X_{mid}$$
$$\underline{r_1 > \max(0, -2p_1 X_{mid})}$$
$$2p_2 X + r_2 < 0 \text{ if } X > X_{mid}$$
$$\underline{r_2 < \min(-2p_2 X_g, -2p_2 X_{mid})}$$

2. *Satisfy the task requirements* (obey task limits, stay at the goal position). Because of the task limits ($\Phi < |\Phi_{max}|, \dot{X} < \left|\dot{X}_{max}\right|$) and since the trolley moves from its start position ($X_0$=0) to the goal position $X_g$:

$$-\dot{X}_{max} < k_1\Phi, k_2\Phi < \dot{X}_{max}$$
$$\underline{-\frac{\dot{X}_{max}}{\Phi_{max}} < k_1, k_2 < \frac{\dot{X}_{max}}{\Phi_{max}}}$$
$$\underline{0 < X_{mid} < X_g}$$

Staying at the goal position requires:

$$0 = \dot{X}_{des}(X_g, 0) = \dot{X}_{mid2} + p_2 X_g^2 + r_2 X_g + k_2 0 =$$
$$= \dot{X}_{mid2} + p_2 X_g^2 + r_2 X_g$$
$$\underline{\dot{X}_{mid2} = -X_g(p_2 X_g + r_2)}$$

3. *Start towards goal.* To ensure that the trolley starts to move toward the goal position $\dot{X}_{des}(0, 0)$ must be positive. This gives the constraint:

$$\dot{X}_{des}(0, 0) = \dot{X}_{mid1} + p_1 0 + r_1 + k_1 0$$
$$\underline{0 < \dot{X}_{mid1} < \dot{X}_{max}}$$

4. *Continuous and fast at $X_{mid}$.* Continuity of $\dot{X}_{des}$ trajectory at $X_{mid}$ is ensured by the constraint $\forall\Phi : \dot{X}_{des}(X_{mid} - \epsilon, \Phi) = \dot{X}_{des}(X_{mid} + \epsilon, \Phi))$. So $\underline{k_1 = k_2}$. Under the condition of continuous trajectory at $X_{mid}$, $\dot{X}_{des}$ is clearly maximal at $X_{mid}$ (after $X_{mid}$ $\dot{X}_{des}$ is decreasing). To achieve the best time, maximal $\dot{X}_{des}$ (that is $\dot{X}_{des}$ at $X_{mid}$) should be as large as possible, that is $\dot{X}_{max}$ (ignoring $\Phi$).

$$\dot{X}_{des}(X_{mid} - \epsilon, 0) = \dot{X}_{des}(X = X_{mid} + \epsilon, 0) = \dot{X}_{max}$$
$$\dot{X}_{mid1} + p_1 {X_{mid}}^2 + r_1 X_{mid} = \dot{X}_{max} =$$
$$= \dot{X}_{mid2} + p_2 {X_{mid}}^2 + r_2 X_{mid}$$
$$\underline{r_1 = \frac{\dot{X}_{max} - \dot{X}_{mid1}}{X_{mid}} - p_1 X_{mid}}$$
$$\dot{X}_{max} = -{X_g}^2 p_2 - X_g r_2 + p_2 {X_{mid}}^2 + r_2 X_{mid}$$
$$\dot{X}_{max} = p_2(X_{mid}^2 - {X_g}^2) + r_2(X_{mid} - X_g)$$
$$\underline{r_2 = \frac{\dot{X}_{max}}{X_{mid} - X_g} - p_2(X_{mid} + X_g)}$$

Considering the constraints on functions derivatives this gives:

$$r_1 = \frac{\dot{X}_{max} - \dot{X}_{mid1}}{X_{mid}} - p_1 X_{mid} > \max(0, -2p_1 X_{mid})$$
$$r_2 = \frac{\dot{X}_{max}}{X_{mid} - X_g} - p_2(X_{mid} + X_g) < \min(-2p_2 X_g, -2p_2 X_{mid})$$

Let us consider all four possible cases:

- $\underline{p_1 \geq 0}$ then $\max(0, -2p_1 X_{mid}) = 0$

$$r_1 = \frac{\dot{X}_{max} - \dot{X}_{mid1}}{X_{mid}} - p_1 X_{mid} > 0$$
$$p_1 < \frac{\dot{X}_{max} - \dot{X}_{mid1}}{X_{mid}^2}$$

- $\underline{p_1 < 0}$ then $\max(0, -2p_1 X_{mid}) = -2p_1 X_{mid}$

$$r_1 = \frac{\dot{X}_{max} - \dot{X}_{mid1}}{X_{mid}} - p_1 X_{mid} > -2p_1 X_{mid}$$
$$p_1 > \frac{-(\dot{X}_{max} - \dot{X}_{mid1}}{X_{mid}^2}$$

- $\underline{p_2 \geq 0}$ then $\min(-2p_2 X_g, -2p_2 X_{mid}) = -120p_2$

$$r_2 = \frac{\dot{X}_{max}}{X_{mid} - X_g} - p_2(X_{mid} + X_g) < -120p_2$$
$$p_2 < \frac{\dot{X}_{max}}{(X_g - X_{mid})^2}$$

- $\underline{p_2 < 0}$ then $\min(-2p_2 X_g, -2p_2 X_{mid}) = -2p_2 X_{mid}$

$$r_2 = \frac{\dot{X}_{max}}{X_{mid} - X_g} - p_2(X_{mid} + X_g) < -2p_2 X_{mid}$$
$$p_2 > \frac{-\dot{X}_{max}}{(X_g - X_{mid})^2}$$

This gives the constraints on $p_1$ and $p_2$:

$$-\frac{\dot{X}_{max} - \dot{X}_{mid1}}{X_{mid}^2} < p_1 < \frac{\dot{X}_{max} - \dot{X}_{mid1}}{X_{mid}^2}$$
$$-\frac{\dot{X}_{max}}{(X_g - X_{mid})^2} < p_2 < \frac{\dot{X}_{max}}{(X_g - X_{mid})^2}$$

5. *The trolley travels fast.* To ensure that the trolley reaches the goal position fast, one additional constraint is needed. This constraint requires that the speed of the trolley is large enough ($\dot{X}_{des} > 0.02\,\dot{X}_{max}$) when the trolley is far from the goal position ($X \leq 0.98\,X_g$):

$$
\begin{aligned}
&\dot{X}_{des}(X,0) > 0.02\,\dot{X}_{max} \text{ if } X \in [X_{mid}, 0.98\,X_g] \\
&\dot{X}_{des}(0.98X_g,0) > 0.02\dot{X}_{max} \text{ (since } \dot{X}_{des} \text{ is dec. in X)} \\
&\dot{X}_{mid2} + (0.98X_g)^2 p_2 + 0.98X_g r_2 > 0.02\dot{X}_{max} \\
&-X_g(p_2X_g + r_2) + (0.98X_g)^2 p_2 + 0.98X_g r_2 > 0.02\dot{X}_{max} \\
&(0.98^2 - 1)X_g p_2 50 - r_2 > \frac{\dot{X}_{max}}{X_g} \\
&-1.98X_g p_2 - \frac{\dot{X}_{max}}{X_{mid} - X_g} + p_2(X_{mid} + X_g) > \frac{\dot{X}_{max}}{X_g} \\
&p_2(-1.98X_g + X_{mid} + X_g) > \frac{\dot{X}_{max}}{X_g} + \frac{\dot{X}_{max}}{X_{mid} - X_g} \\
&\underline{p_2 < \frac{\frac{\dot{X}_{max}}{X_g} + \frac{\dot{X}_{max}}{X_{mid}-X_g}}{(X_{mid} - 0.98X_g)}}
\end{aligned}
$$

To summarize the derivation of the constraints:

- Quantitative strategies $S_{q1}$ have the first four properties, giving the constraints on qualitative parameters:

$$
\begin{aligned}
&0 < X_{mid} < X_g \text{ , } 0 < \dot{X}_{mid1} < \dot{X}_{max} \\
&-\frac{\dot{X}_{max}}{\Phi_{max}} < k_1 = k_2 < \frac{\dot{X}_{max}}{\Phi_{max}} \\
&-\frac{\dot{X}_{max} - \dot{X}_{mid1}}{X_{mid}^2} < p_1 < \frac{\dot{X}_{max} - \dot{X}_{mid1}}{X_{mid}^2} \\
&-\frac{\dot{X}_{max}}{(X_g - X_{mid})^2} < p_2 < \frac{\dot{X}_{max}}{(X_g - X_{mid})^2} \\
&r_1 = \frac{\dot{X}_{max} - \dot{X}_{mid1}}{X_{mid}} - p_1 X_{mid} \\
&r_2 = \frac{\dot{X}_{max}}{X_{mid} - X_g} - p_2(X_{mid} + X_g) \\
&\dot{X}_{mid2} = -X_g(X_g p_2 + r_2)
\end{aligned}
$$

- Quantitative strategies $S_{q2} \subset S_{q1}$ also have the fifth property, giving additional constraint:

$$
p_2 < \frac{\frac{\dot{X}_{max}}{X_g} + \frac{\dot{X}_{max}}{X_{mid}-X_g}}{(X_{mid} - 0.98X_g)} \tag{4.11}
$$

## 4.3 Skill Reconstruction in Acrobot Domain

The reconstruction of human swing-up control skill follows the approach described in Section 4.1 with details specific to the acrobot domain. This domain is described in Section 2.3.2.

First we give details of learning the operator's swing-up strategy and present an example of the induced strategy. This quantitative strategy is then abstracted to a qualitative strategy and qualitative control rules that are used to explain the operator's control skill. In Section 4.2.3 we use the qualitative strategy as a space for controller optimization by transforming it into a set of quantitative strategies.

### 4.3.1 Learning Quantitative Swing-Up Strategy

First, the generalized operator's trajectory is induced from an execution trace. We choose the dependent trajectory variable to be $q2$ according to our notion of the causality in the system, namely that the control torque is applied in the $q2$ joint. The choice of dependent variables is discusses in Chapter 7. It is sufficient that the generalized trajectory determines the desired $q2$ as a function of the current state. It is not necessary that this trajectory also determines the $q2$ velocity. If we are able to compute the desired $q2$ angle as a function of the state, and we know the time step $dt$ after which the desired $q2$ angle is to be achieved, the desired $q2$ velocity is defined by the current $q2$ angle and velocity: $\dot{q2}$ that achieves $q2_{des}$ after time $dt$ is $(q2_{des} - q2)/dt$, assuming that $\dot{q2}$ is constant over time step $dt$.

GoldHorn is used to induce constraints between $q2$ and other state variables from the execution trace. The most significant (according to the measures given in Section 2.4.3) among the induced equations are used to express the desired $q2$ as a function of the current state. There are cases when a single equation does not uniquely define the desired $q2$ angle (for example, the sign of $q2$ angle is undefined). In such a case it is possible to consider other equations as all of them reflect the constraints between the state variables along the operator's trajectory. One such example of combining equations is presented in the next section.

When the trajectory is known, the action which achieves the desired next state on the trajectory is computed using a locally linear model learned by LWR from some execution traces, as described in Section 4.1.1. At each time step the controller considers the current state and computes $q2_{des}$ according to the generalized trajectory. Afterwards it learns a model of system dynamics near the current state by LWR and uses this approximate model to select torque control aiming to correct the constraint error, i.e. the difference between $q2_{des}$ and the current value of $q2$. Precisely, LWR learns local linear model near the current state, i.e. $\dot{q2}$ at next times step as a linear function of $q1$, $\dot{q1}$, $q2$ and the torque $\tau$ near the current state:

$$\dot{q2}(t + dt) = k_1 q1(t) + k_2 \dot{q1}(t) + k_3 q2(t) + k_4 \dot{q2}(t) + k_5 \tau(t) + c$$

Then we would like to take an action which corrects the constraint error. The user-defined parameter $s$, set to 3 in all our experiments, determines how many time steps are allowed to correct this error. When $s$=1 the desired $q2_{des}$ is achieved after the next time step; if $s$=3 the desired $q2_{des}$ is to be achieved in three time steps. This assumes, of course, that $q2_{des}$ stays constant over $s$ steps. However, as the dynamics will be changing (different local linear models), this is only an approximation. The desired $\dot{q2}_{des}$, aiming to correct this error, can be approximated by:

$$\dot{q2}_{des}(t + dt) = \frac{q2_{des}(t) - q2(t)}{dt\ s}$$

The corresponding torque can be expressed as:

$$\tau(t) = \frac{\dot{q2}_{des}(t{+}dt) - k_1 q1(t) - k_2 \dot{q1}(t) - k_3 q2(t) - k_4 \dot{q2}(t) - c}{k_5}$$

Note that this approach is slightly different than with experiments in the crane domain (Section 4.2.1). Here the generalized trajectory is induced as the constraints among the variables in the same state, whereas in the crane domain we induced the constraints between the current and the next state. This difference is discussed is Section 4.1.1.

This approach to skill reconstruction turned out to be generally very effective. For most of the successful execution traces that we experimented with, GoldHorn induced a generalized trajectory which allows the controller to complete the swing-up task. However, the time needed to complete the task varied significantly with the execution trace and the particular differential equation used. We found that differential equations with linear and quadratic factors were usually most successful. Terms of higher orders usually did not improve the performance. Another interesting, but not surprising, observation was that usually a more successful trajectory was induced from a faster execution trace. So individual differences between the operators were reflected in the generalized trajectories induced from the operator's traces. In the following paragraph we present the strategy learned from the most successful operator's execution trace.

### Example of Induced Swing-Up Strategy

Program GoldHorn was used to induce differential equations which describe constraints between variables $q1$, $\dot{q1}$, $qcm$ and $q2$ in the most successful execution trace. The choice of the trajectory variables and the selection of equations is also discussed in Chapter 7.

The following differential equation was the best rated equation found by GoldHorn (disregarding useless or meaningless equations such as trivial identities also found by GoldHorn):

$$0.947 + 0.258q1^2 + 0.032q1\dot{q1} - 0.006\dot{q1}^2 - q2^2 = 0$$

Here $q2_{des}$ can be expressed as:

$$q2_{des} = \pm\sqrt{0.947 + 0.258q1^2 + 0.032q1\dot{q1} - 0.006\dot{q1}^2} \tag{4.12}$$

But there are two problems with this equation: first, the sign of the square root is not known, and second, the square root does not have real solutions when the argument of the square root is negative. This happens just inside a small region near the edge of the state space. It reflects the fact that, when searching for equations with good fit GoldHorn does not know that $q2$ must always be real-valued. Let us call the region of the state space where the Equation 4.12 has real solutions the *main region*, and the rest of the state space the *edge region*. In order to determine $q2_{des}$ at the edge region and the sign of $q2_{des}$ in the main region of the state space, we can consider another equation found by GoldHorn:

$$0.060 - 0.754q2 - qcm = 0$$

This determines the sign of $q2_{des}$ and $q2_{des}$ as follows:

$$\begin{aligned} q2_{des} &= (0.060 - qcm)/0.754 \\ sign(q2_{des}) &= sign(0.060 - qcm) \end{aligned} \tag{4.13}$$

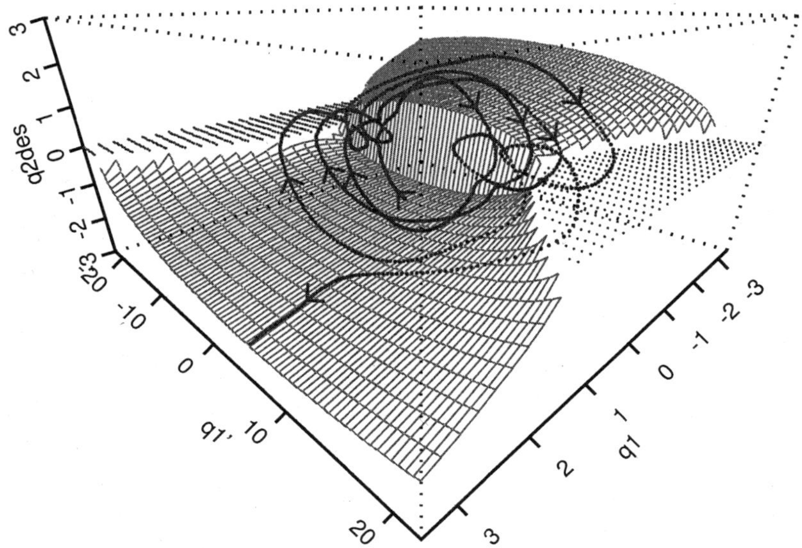


Figure 4.5: Swing-up strategy in the main region: the surface with solid lines is the desired $q2$ (on the $z$-axis) for given $q1$ and $\dot{q1}$. To represent $q2_{des}$ on a 3-D plot, the sign of $q2_{des}$ is approximated by $-sign(q1)$ (in the trace $qcm$ changes its sign near $q1 = 0$). The dotted surface at $q2_{des}$=0 is the edge region of the state space. The dotted curve represent the actual $q2$ in the original trace and start after the first maximum of $q1$. The arrows indicate the time order.

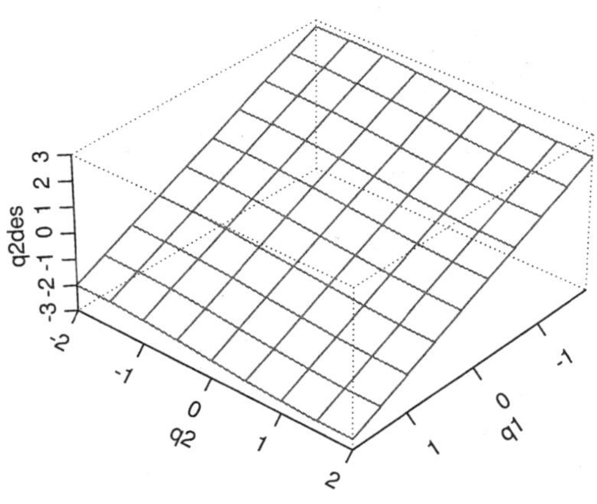


Figure 4.6: Swing-up strategy in the edge region of the state space (where equation 4.12 has no real solutions): $x$ and $y$-axis represent $q1$ and $q2$ angle and $z$-axis represents $q2_{des}$ (Equation 4.13) for $qcm$ angle computed from the corresponding $q1$ and $q2$ values. The range of $q1$ is only from -1.9 rad. to 1.9 rad. because when $|q1| > 1.9$ rad. the Equation 4.13 is never considered.

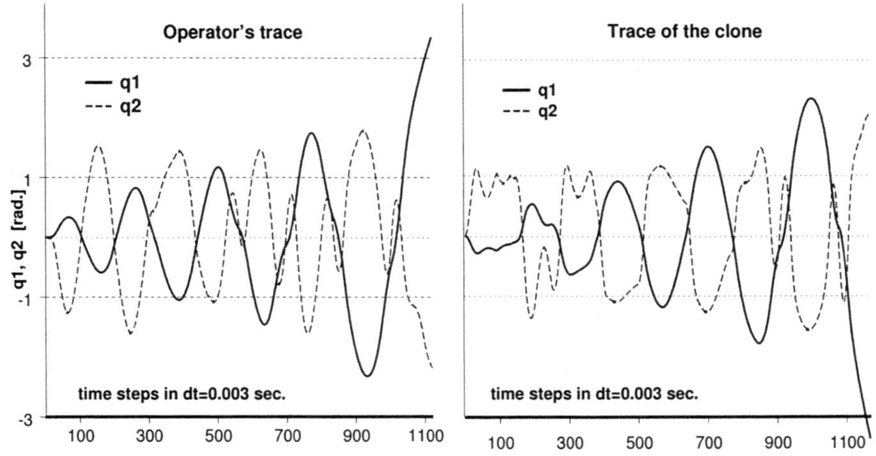


Figure 4.7: Operator's trace and its clone: on the left is the most successful operator's execution trace, which completes the task in 1124 steps, and on the right is the trace of its clone, which completes the task in 1169 steps. Operator and its clone start in the opposite direction, since the value 0.060 in Equation 4.13 is a chance numerical approximation of 0.

Here, 0.060 is a chance numerical approximation of 0. It is interesting that Equation 4.13 alone defines a very simple successful strategy, which requires about 4000 steps to conclude the swing-up task. This strategy is not very efficient since it only roughly reflects the operator's trajectory (its error estimate is much higher than the error estimate of Equation 4.12), but it gives a very good insight in what the operator is trying to do.

Both Equations 4.12 and 4.13 reflect the constraints between the state variables in the execution trace; so we can consider both to compute the desired $q2$ angle. Here it is important to note that the behaviour of $q2_{des}$ is qualitatively similar in both regions of the state space. $q2_{des}$ is in both cases negatively proportional to $q1$ (see Figure 4.5 and Figure 4.6 for details).

Since Equation 4.12 has lower error estimate, we use Equation 4.13 only to determine the sign of $q2_{des}$ in the main region, and to determine also the absolute value of $q2_{des}$ in the edge region. Combining both equations in this way we get the desired $q2$ as:

$$\begin{aligned}
&\text{if } 0.947 + 0.258q1^2 + 0.032q1\dot{q1} - 0.006\dot{q1}^2 > 0 \text{ then } \quad \{\text{main region}\} \\
&\quad q2_{des} = sign(0.060 - qcm)\sqrt{0.947 + 0.258q1^2 + 0.032q1\dot{q1} - 0.006\dot{q1}^2} \\
&\text{else} \quad \{\text{edge region}\} \\
&\quad q2_{des} = (0.060 - qcm)/0.754
\end{aligned} \tag{4.14}$$

This Rule defines a quantitative control strategy and was used to control the acrobot by computing the desired $q2$ first and then the control torque as described in the previous section. The resulting control depended on the quantity of data, i.e. the number of execution traces used to learn the system's dynamics with locally weighted linear regression. When the only data used to learn the system's dynamics was the execution trace used to induce the $q2_{des}$ trajectory, it took about 1500 time steps to complete the swing-up task. To obtain a more accurate system dynamics model, we used more execution traces as data for LWR. As the knowledge of the system's dynamics got better, the time needed to complete the task got shorter. The improvements due to greater number of learning cases resulted in execution time of about 1160 time steps. After reaching this level of performance, the addition of further learning cases did not improve the performance any further. Figure 4.7 presents the original trace and the trace of the clone where four traces were used to learn the system's dynamics. The clone completes the task in 1169 steps which is slightly worse than the original trace but much better than any other human execution trace.

#### Notes on Experiments

In the presented example of the induced strategy we also used the center of mass angle ($qcm$) as a state variable in the execution trace. $qcm$ can be expressed as a function of $q1$ and $q2$, but we used it because it gives very important and easy-to-understand information about the acrobot position. We wanted the induced strategies to be close to the operators' thinking about their skills. When the operators described their swing-up skill (Blažina et al., 1996; Miletić, 1997), one of the fastest operators (operator E.) claimed that he takes the control actions just according to the values of $q2$ and $qcm$. As noted in the previous section, the constraint between $qcm$ and $q2$ suffices for a simple, although not very efficient swing-up strategy defined by Equation 4.13. The strategy defined by this equation is similar to a simple swing-up strategy described in Section 2.3.2.

We used traces sampled at the sample time $dt = 0.003$ sec. Since a successful trace of a skilled operator lasts about 1500 time steps, this gives about 1500 learning examples for the GoldHorn program.

### 4.3.2 Qualitative Swing-Up Strategy

#### Qualitative Abstraction of Quantitative Strategy

The induced quantitative strategy (Rule 4.14) can be abstracted to a qualitative strategy as described in Section 4.1.2. We used qualitative abstraction to derive qualitative strategy in the main region of state space from the induced quantitative strategy. Since the behaviour of the quantitative strategy is qualitatively similar in both regions of the state space, the qualitative strategy from the main region can also be used in the edge region of the state space. The derivation is given in Section 4.3.4 and results in a qualitative control strategy:

$$\begin{array}{l}
q2Sign = sign(smallQcm - qcm) \\
\text{if } sign(q1) = sign(\dot{q}_1) \text{ then } \quad \{\text{swing-up}\} \\
\quad \text{if } q1 \textit{ near local extreme} \text{ then} \\
\qquad q2_{des} = M^{+,+}(|q1|, |\dot{q1}|) \times q2Sign \\
\quad \text{else } q2_{des} = M^{+,-}(|q1|, |\dot{q1}|) \times q2Sign \\
\text{else } \quad \{sign(q1) \neq sign(\dot{q1}), \text{swing-down}\} \\
\quad \text{if } q1 \textit{ near zero} \text{ then} \\
\qquad q2_{des} = M^{-,-}(|q1|, |\dot{q1}|) \times q2Sign \\
\quad \text{else } q2_{des} = M^{+,-}(|q1|, |\dot{q1}|) \times q2Sign
\end{array} \tag{4.15}$$

Here *smallQcm*, $q1$*near zero* and $q1$*near local extreme* are *qualitative terms*. They are just names we gave to the constraints derived by qualitative abstraction:

$$\begin{aligned} q1 \textit{ near zero } &\Leftrightarrow |q1| < 0.062 \left|\dot{q1}\right| \\ q1 \textit{ near local extreme } &\Leftrightarrow |q1| > 0.375 \left|\dot{q1}\right| \\ smallQcm &= 0.060 \end{aligned} \tag{4.16}$$

It is interesting that this constraints indeed correspond to events that might be important for the operator, as for example acrobot swinging through the vertical or reaching the highest point of the current swing.

### Explanation of Qualitative Strategy

The qualitative strategy given by Rule 4.15 can also be turned into qualitative control rules that can be used to explain the operator's control skill. A qualitative control rule is simply a qualitative constraint and the condition under which the qualitative constraint applies.

Understanding qualitative rules requires some knowledge of the swing-up task. As described in Section 2.3.2, every possible solution to the swing-up task involves periodic oscillations of $q1$ such that the amplitude of $q1$ increases in consecutive cycles. This means that the solution requires several cycles, where each cycle consists of two *swing-up* parts (when $sign(q1) = sign(\dot{q1})$) and two *swing-down* parts (when $sign(q1) \neq sign(\dot{q1})$). Increasing $|q2|$ corresponds to *shortening* of the acrobot and decreasing $|q2|$ to *lengthening* of the acrobot.

The qualitative strategy from Rule 4.15 gives five qualitative control rules:

1. *The $q2$ sign rule*: $sign(q2) = sign(smallQcm - qcm)$; the sign of $q2$ is the opposite of the sign of the approximate center of mass angle and the sign of $q2$ changes when the center of mass angle is near zero.

2. *The $q2$ magnitude rule 1*: $|q2_{des}|$ is positively related to $|q1|$ if $sign(q1) = sign(\dot{q1})$ or $q1$ is not *near zero*. During the swing-up ($sign(q1) = sign(\dot{q1})$) $|q1|$ is increasing, so this rule causes increasing of $|q2_{des}|$, i.e. shortening of the acrobot. During the swing-down when $q1$ is not *near zero* and $|q1|$ is decreasing, this rule causes lengthening of the acrobot. An important consequence of this rule is, that the amplitude of $|q2_{des}|$ increases in consecutive cycles since the $|q1|$ amplitude increases in consecutive cycles.

3. *The $q2$ magnitude rule 2*: $|q2_{des}|$ is negatively related to $\left|\dot{q1}\right|$ if $sign(q1) \neq sign(\dot{q1})$ or $q1$ is not *near local extreme*. During the swing-down ($sign(q1) \neq sign(\dot{q1})$) $\left|\dot{q1}\right|$ is increasing, so this causes lengthening of the acrobot. If during the swing-down, when $q1$ is not *near zero*, $\left|\dot{q1}\right|$ is decreasing, this rule causes shortening of the acrobot. During the swing-up ($\left|\dot{q1}\right|$ is decreasing), when $q1$ is not *near local extreme*, this rule causes shortening of the acrobot, similar as the $q2$ magnitude rule 1. Another important consequence of this and the previous rule is that $|q2_{des}|$ is the same if $|q1|$ and $\left|\dot{q1}\right|$ both increase (in two consecutive cycles).

4. *The swing-up-end lengthening rule*: $|q2_{des}|$ is positively related to $\left|\dot{q1}\right|$ if $sign(q1) = sign(\dot{q1})$ and $q1$ *near its local extreme*. This rule causes lengthening of the acrobot at the end of the swing-up since $\left|\dot{q1}\right|$ is decreasing.

5. *The swing-down-end shortening rule*: $|q2_{des}|$ is negatively related to $|q1|$ if $sign(q1) \neq sign(\dot{q1})$ and $q1$ *near zero*. This rule causes shortening of the acrobot at the end of the swing-down and tells that the transition from lengthening to shortening (i.e. $q2$ crossing zero) must happen at the end of the swing-down cycle. This rule is similar to the $q2$ sign rule. However, it only tells when to change the sign of $q2$, but does not specify whether $q2$ is positive or negative.

We summarize these control rules as:

- The acrobot is to be controlled so that the sign of $q2$ is the opposite of the sign of the center of mass angle and the sign of $q2$ changes when the center of mass angle is near zero.
- Lengthening must take place at the end of the swing-up cycle (when $|q1|$ is approaching its local maximum) and during the swing-down cycle, except at the end of the swing-down cycle, when $|q1|$ approaches zero. Otherwise shortening must take place. This causes $q2$ to reach its local extreme before the end of the swing-up, while $|q1|$ is still increasing.
- The amplitude of $|q2|$ increases in consecutive cycles. The same $|q2|$ (in two consecutive cycles) is achieved at bigger $|q1|$ and bigger $|\dot{q1}|$.

Considering the only possible solution to the swing-up task involves periodic oscillations of $qcm$ so that the amplitude of $qcm$ increases in consecutive cycles, and that each cycle consists of two swing-up and two swing-down parts, the induced qualitative rules allow us to describe the operator's strategy:

Initially ($q1 = q2 = \dot{q}_1 = \dot{q}_2 = 0$) increase $q2$ (see Rule 4.14); this causes the decrease of $q1$ so the swing-up part of the cycle begins. Keep increasing $q2$ (shortening by the $q2$ magnitude rules). At the end of the swing-up decrease $q2$ by the swing-up-end lengthening rule. After $q1$ reaches its negative extreme, swing-down part of the cycle begins, so keep $q2$ decreasing (lengthening by the $q2$ magnitude rule 1). $qcm$ is now negative and increasing, and $q2$ is still positive (the $q2$ sign rule) and decreasing. At the end of the swing-down or when $qcm$ reaches 0.060 rad (whatever happens first), $q2$ crosses zero by the $q2$ sign rule and the swing-down-end shortening rule. Now swing-up begins: $q1$ is positive and increasing, and $q2$ is negative and decreasing. Keep decreasing $q2$ (shortening by the $q2$ magnitude rules). At the end of the swing-up increase $q2$ by the swing-up-end lengthening rule. After $q1$ reaches its positive extreme, swing-down begins, so keep increasing $q2$ (lengthening by the $q2$ magnitude rule 2). At the end of the swing-down $q2$ crosses zero and shortening begins by the swing-down-end shortening rule. Now a new cycle begins.

The main idea of this strategy is very simple: drive the second link so as to inject energy in the phase with the swinging of the first link, similar like a child pumping a swing. However, the nonlinear relation between $q2$ and the state variables is very important. The initial swing of the acrobot is a huge problem for symmetric swing-up strategies like the ATAN strategy (Spong, 1994) since they require many small swing cycles at the beginning. Efficient initial cycle is also hard to learn for human operators. After the first $q2$ deflection exact timing is very important. The induced sign rule and the nonlinear magnitude rules solve this problem efficiently. The induced lengthening rule, which improves the time needed to complete the task is very interesting. It also has a very good explanation: by lengthening the acrobot at the end of the swing-up the potential energy of the system is increased. This rule is a very

Table 4.1: Values of qualitative parameters used to transform qualitative strategy given by Rule 4.15 into quantitative strategies.

| Qualitative parameter | Values of the qualitative parameter |
|---|---|
| $smallQcm$ | $0.0, 0.05, 0.1$ |
| $q1$ *near zero* | $\|q1\| < k_1 \left\|\dot{q1}\right\|, k_1 = 0.05, 0.1, 0.15$ |
| $q1$ *near local extreme* | $\|q1\| > k_2 \left\|\dot{q1}\right\|, k_2 = 0.2, 0.4, 0.6$ |
| $f_1^{+,+}, f_2^{+,-}, f_3^{-,-}, f_4^{+,-}$ | randomly generated piece-wise linear QCFs |

important part of the experienced operator's swing-up strategy and cannot be observed in the simple "beginners" strategies.

### 4.3.3 Transforming Qualitative Strategy into Quantitative Strategies

The qualitative strategy given by Rule 4.15 defines a set of quantitative strategies with free qualitative parameters. These are the qualitative terms $smallQcm$, $q1$ *near zero*, $q1$ *near local extreme* and four qualitatively constrained functions. To transform the qualitative strategy into executable quantitative strategies this qualitative parameters are to be concretised into numerical values or real-valued functions. When transforming the qualitative strategy into quantitative strategies we use the domain knowledge consisting of the control task limits ($|q1| \leq \pi$, $\left|\dot{q1}\right| \leq 20$, $|q2| \leq q2_{max}$) and the requirement that $q2_{des}$ is a continuous function. We also used some knowledge about "reasonable" values of qualitative parameters. For example, qualitative term $smallQcm$ was concretised into values 0.0, 0.05 or 0.1, which is close to the value of $smallQcm$ from the induced quantitative strategy.

Using the domain knowledge about the task limits we define functions $f_i^{s_1,s_2}$ as arbitrary functions mapping $[-\pi, \ldots, \pi] \times [0, \ldots, 20] \mapsto [0, \ldots, q2_{max}]$. and respecting the qualitative constraints given by signs $s_1$ and $s_2$. Then the qualitative strategy (Rule 4.15) defines a set of quantitative strategies:

$$
\begin{aligned}
&q2Sign = sign(smallQcm - qcm) \\
&\text{if } sign(q1) = sign(\dot{q}_1) \text{ then } \quad \{\text{swing-up}\} \\
&\quad \text{if } q1 \textit{ near local extreme} \text{ then} \\
&\qquad q2_{des} = f_1^{+,+}(|q1|, \left|\dot{q1}\right|) \times q2Sign \\
&\quad \text{else } q2_{des} = f_2^{+,-}(|q1|, \left|\dot{q1}\right|) \times q2Sign \\
&\text{else } \quad \{sign(q1) \neq sign(\dot{q1}), \text{swing-down}\} \\
&\quad \text{if } q1 \textit{ near zero} \text{ then} \\
&\qquad q2_{des} = f_3^{-,-}(|q1|, \left|\dot{q1}\right|) \times q2Sign \\
&\quad \text{else } q2_{des} = f_4^{+,-}(|q1|, \left|\dot{q1}\right|) \times q2Sign
\end{aligned}
\tag{4.17}
$$

To achieve continuous $q2_{des}$ function, the following constraints are to be satisfied:

$$
\begin{aligned}
&f_2^{+,-}(k_2\dot{q1}, \dot{q1}) = M_1^{+,+}(k_2\dot{q1}, \dot{q1}) \\
&f_2^{+,-}(0, \dot{q1}) = M_3^{-,-}(0, \dot{q1}) \\
&f_4^{+,-}(q1, 0) = M_1^{+,+}(q1, 0) \\
&f_4^{+,-}(k_1\dot{q1}, \dot{q1}) = M_3^{-,-}(k_1\dot{q1}, \dot{q1})
\end{aligned}
$$

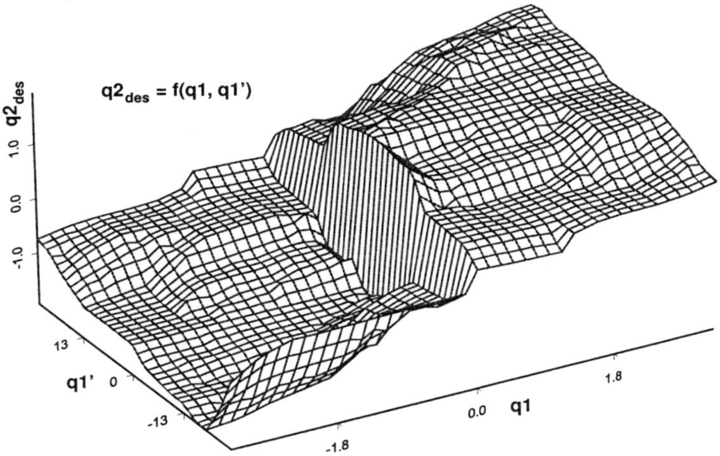


Figure 4.8: Quantitative strategy with values of qualitative parameters $smallQcm = 0$, $k_1 = 0.05$ and $k_2 = 0.2$ and randomly generated QCFs. On $z$-axis is $q2_{des}$ for the corresponding $q1$ ($x$-axis) and $\dot{q1}$ ($y$-axis). The quantitative strategy needs 846 time steps to successfully conclude the swing-up task. To represent the strategy on a 3-D plot, $qcm$ is supposed to cross zero at $q1 = 0$.

Quantitative strategies defined by Rule 4.17 and respecting the above constraints were tested on the simulator. To transform QCFs into real-valued functions we used the procedure described in Section 4.1.3 to generate piece-wise linear QCFs at random. The QCFs were generated with $C = 8 \times 8$ ($9 \times 9$ cells for each function). Qualitative terms $q1$ *near zero*, $q1$ *near local extreme* and $smallQcm$ were each assigned 3 different values or functions. The values of qualitative parameters are given in Table 4.1. For each set of qualitative term's value we generated 200 different sets of functions $f_1^{+,+}, f_2^{+,-}, f_3^{-,-}, f_4^{+,-}$. This gives $3 \times 3 \times 200 = 1800$ different quantitative strategies.

738 out of 1800 different quantitative strategies completed the task in 2000 steps or less, with mean time 1308 $dt$ steps. Many strategies were faster than 1000 steps, which is considerably better than any human operator. Note that the operators were considered successful even if they completed the task in 10000 steps. A typical successful trace lasted 2000 steps. The best strategy (see Figure 4.8) needed only 846 $dt$ steps to conclude the swing-up task. Note that the unsuccessful strategies failed to complete the task mainly because $q2$ angle exceeded the maximum allowed $q2_{max} = 2.2rad.$. Although the desired $q2_{des}$ is always smaller than $q2_{max}$ the actual $q2$ exceeded $q2_{max}$ because the maximum torque is too small to reach the exact $q2_{des}$. When the experiment was repeated with $q2_{max} = 2.5rad.$, 1571 out of 1800 strategies completed the task in 2000 steps or less.

### 4.3.4 Derivation of Qualitative Swing-Up Strategy

In Section 4.3.1 we induced an operator's quantitative strategy (Rule 4.14) for the acrobot swing-up task. Here we derive the qualitative strategy in the main region (where Equation 4.12 has real solutions) from the induced quantitative strategy.

For the simplicity of the derivation, we use a similar notation as used in (Forbus, 1984) to

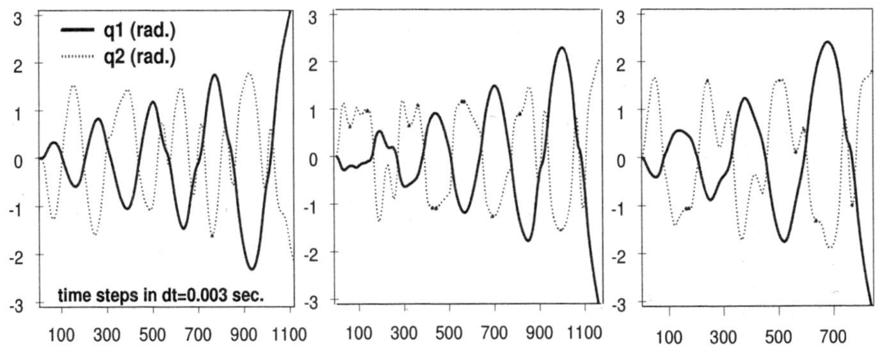


Figure 4.9: Traces of 3 strategies: on the left is the original operator's trace, in the middle is the trace of the induced quantitative strategy (Rule 4.14), on the right is the trace of the quantitative strategy obtained from the qualitative strategy with values of qualitative parameters $smallQcm = 0$, $k_1 = 0.05$ and $k_2 = 0.2$ and randomly generated qualitative functions. All the traces share the same qualitative strategy. The first two traces need 1124 and 1169 time steps to conclude the swing-up task. The third trace is much faster and concludes the task in 846 steps.

indicate that one variable is positively or negatively related to another variable. The notation $y \propto_+ x$ means that $y$ is positively related to $x$. Similarly $y \propto_- x$ means that $y$ is negatively related to $x$.

The first obvious rule is:

$$sign(q2_{des}) = sign(0.060 - qcm)$$

Since the value $qcm = 0.060$ is a small angle, we abstract this value to the qualitative term $smallQcm$. To find qualitative constraints on $|q2_{des}|$ we define $q2S = |q2_{des}|^2$. Then Equation 4.12 can be written as:

$$q2S = 0.947 + 0.258q1^2 + 0.032q1\dot{q1} - 0.006\dot{q1}^2$$

To find qualitative constraints we compute partial derivatives. These are:

$$\frac{\partial q2S}{\partial q1} = 0.516q1 + 0.032\dot{q1} > 0 \Leftrightarrow q1 > -0.062\dot{q1}$$
$$\frac{\partial q2S}{\partial \dot{q1}} = 0.032q1 - 0.012\dot{q1} > 0 \Leftrightarrow q1 > 0.375\dot{q1}$$

Since $|\dot{q1}| < 20$, but $|q1| < \pi$, we abstract the two constraints to qualitative terms $q1$ *near zero* and $q1$ *near local extreme*:

$$|q1| < 0.062\,|\dot{q1}| \Leftrightarrow \text{q1 } \textit{near zero}$$
$$|q1| > 0.375\,|\dot{q1}| \Leftrightarrow \text{q1 } \textit{near local extreme}$$

Let us consider all four possible cases:

- $q1 > 0, \dot{q1} > 0$: $q2S \propto_+ q1$ since $\frac{\partial q2S}{\partial q1} > 0$. $\frac{\partial q2S}{\partial \dot{q1}} > 0$ is true when $q1 > 0.375\dot{q1}$. Therefore $q2S \propto_+ \dot{q1}$ when $q1$ *near local extreme*, otherwise $q2S \propto_- \dot{q1}$.
- $q1 < 0, \dot{q1} < 0$: $q2S \propto_- q1$ since $\frac{\partial q2S_{des}}{\partial q1} < 0$. $\frac{dq2S}{d\dot{q1}} > 0$ is true when $q1 > 0.375\dot{q1}$, that is when $|q1| < 0.375\left|\dot{q1}\right|$, i.e. when not $q1$ *near local extreme*. Therefore $q2S \propto_- \dot{q1}$ when $q1$ *near local extreme*, otherwise $q2S \propto_+ \dot{q1}$.
- $q1 > 0, \dot{q1} < 0$: $q2S \propto_+ \dot{q1}$. $\frac{\partial q2S}{\partial q1} > 0$ when $q1 > -0.062\dot{q1}$, that when not $q1$ *near zero*. Therefore $q2S \propto_- q1$ when $q1$ is *near zero*, otherwise $q2S \propto_+ q1$.
- $q1 < 0, \dot{q1} > 0$: $q2S \propto_- \dot{q1}$. $q2S \propto_+ |q1|$ when $q1$ *near zero*, otherwise $q2S \propto_- q1$.

Let us summarize all four cases:

1. $sign(q1) = sign(\dot{q1})$ then $q2S \propto_+ |q1|$ and $q2S \propto_+ \left|\dot{q1}\right|$ if $q1$ is *near local extreme*, $q2S \propto_- \left|\dot{q1}\right|$ otherwise.
2. $sign(q1) \neq sign(\dot{q1})$ then $q2S \propto_- \left|\dot{q1}\right|$ and $q2S \propto_- |q1|$ if $q1$ is *near zero*, $q2S \propto_+ |q1|$ otherwise.

Since any qualitative proprotionality applying to $q2S = |q2_{des}|^2$ also applies to $|q2_{des}|$, the qualitative strategy in the main region of the state space is:

$$
\begin{aligned}
& q2Sign = sign(smallQcm - qcm) \\
& \text{if } sign(q1) = sign(\dot{q}_1) \text{ then } \quad \{\text{swing-up}\} \\
& \quad \text{if } q1 \textit{ near local extreme} \text{ then} \\
& \qquad q2_{des} = M^{+,+}(|q1|, |\dot{q1}|) \times q2Sign \\
& \quad \text{else } q2_{des} = M^{+,-}(|q1|, |\dot{q1}|) \times q2Sign \\
& \text{else} \quad \{sign(q1) \neq sign(\dot{q1}), \text{swing-down}\} \\
& \quad \text{if } q1 \textit{ near zero} \text{ then} \\
& \qquad q2_{des} = M^{-,-}(|q1|, |\dot{q1}|) \times q2Sign \\
& \quad \text{else } q2_{des} = M^{+,-}(|q1|, |\dot{q1}|) \times q2Sign
\end{aligned}
\tag{4.18}
$$

By giving names to the derived qualitative constraints, we can write the following qualitative rules:

- *The q2 sign rule*: $sign(q2_{des}) = sign(0.060 - qcm)$
- *The q2 magnitude rule 1*: $|q2_{des}| \propto_+ |q1|$ if $sign(q1) = sign(\dot{q1})$ or $q1$ is not *near zero*.
- *The q2 magnitude rule 2*: $|q2_{des}| \propto_- \left|\dot{q1}\right|$ if $sign(q1) \neq sign(\dot{q1})$ or $q1$ is not *near local extreme*.
- *The swing-up-end lengthening rule*: $|q2_{des}| \propto_+ \left|\dot{q1}\right|$ if $sign(q1) = sign(\dot{q1})$ and $q1$ is *near local extreme*.
- *The swing-down-end shortening rule*: $|q2_{des}| \propto_- |q1|$ if $sign(q1) \neq sign(\dot{q1})$ and $q1$ is *near zero*.

## 4.4 Discussion and Conclusions

### 4.4.1 Discussion

The presented approach extends the idea of problem decomposition for behavioural cloning from the previous chapter and opens new perspectives to the reconstruction of human control skill such as qualitative treatment and optimizing operator's control strategy. Some specific points of interest are:

1. To understand the operator's control strategy, it is important to identify the subgoals that the operator is pursuing at various times. A method for identifying discrete subgoals was developed in (Šuc and Bratko, 1997b). The generalized operator's trajectory can be viewed as a *continuously changing subgoal* and enables skill reconstruction also in domains where the operator's strategy cannot be expressed in terms of a few discrete subgoals.

2. We use methods of *symbolic* machine learning to generalize the operators' trajectories. In principle, any approach to learning could be used to generalize trajectories, including subsymbolic learning such as artificial neural networks. However, using a symbolic approach has the advantage that the resulting generalized trajectory is expressed in an explicit symbolic form and therefore easier to interpret, analyze or correct by considering the knowledge of the control task. One example of such analysis is given in the experiments in the crane domain. We used the fact that the rope length at the goal position $X_g$ should be $L_g$ to correct the learned $L_{des}$ trajectory. The domain knowledge use is discussed in depth in Chapter 7.

3. Because the strategy as a generalized operator's trajectory plus system's dynamics is goal directed, it is robust with respect to small changes in the control task. The experiments in the crane domain show a significant improvement in the robustness of controllers using the generalized operator's trajectory over the controllers in the form of decision or regression trees. Since the system's dynamics can be updated online, strategies are robust with respect to change in the system's dynamics. Of course, this assumes that the operator's trajectory is still appropriate for the changed dynamics of the system.

4. Given a qualitative control strategy, we can tune a controller within the corresponding qualitative constraints. In this way a qualitative control strategy defines an optimization space for better performance. We can also use some background knowledge and derive constraints on the tuned qualitative parameters as demonstrated in the crane domain. By tuning (giving the qualitative parameters real, numerical values) we can achieve better performance than that of the source quantitative strategy.

5. By using a learned qualitative strategy and considering knowledge of the control task to find constraints on qualitative parameters, the learned strategy can be used for considerably different tasks (different goal position, different safety requirements, different parameters of the simulator). For example, we could use the learned qualitative strategy to move the load (possibly different) backwards from the goal position to the start position with more rigorous safety requirements (for example $\dot{X} < 1$m/s and $\Phi < 6$ deg.).

6. When experimenting in the acrobot domain we induced the equation applicable in the main region:

$$q2_{des} = \sqrt{0.947 + 0.258q1^2 + 0.032q1\dot{q1} - 0.006\dot{q1}^2}$$

Using this equation we can derive all the given qualitative rules except the $q2$ sign rule. The sign of $q2$ is not known, but other rules define when $q2$ changes its sign and therefore define two possible qualitative strategies. One idea would be to use qualitative simulation using qualitative system dynamics to prove that one of these two strategies cannot be successful (when $sign(q2) = sign(qcm)$), and the other one (when $sign(q2) = -sign(qcm)$) can be successful. So, qualitative rules and qualitative simulation could be used to reduce the space of possible quantitative strategies and/or to complete an incompletely defined strategy.

### 4.4.2 Conclusions

The presented approach extends the idea of problem decomposition for behavioural cloning from the previous chapter. The approach involves learning separately (a) the system's dynamics and (b) the generalized trajectory the operator is trying to follow. The system's dynamics is learned using locally weighted regression. The trajectory is learned in the form of differential equations. Once the trajectory is known, a controller can easily be constructed since we know the desired next state and a local linear model near the current state. The induced quantitative control strategy often concludes the task faster than the original operator's traces. The induced generalized operator's trajectory provides an insight into the human's subcognitive control skill.

The success of qualitative control strategies in the crane and acrobot domain is the motivation to induce qualitative strategies directly from the execution traces. This idea is explored later in Chapter 6.

# Chapter 5

# Qualitative Induction

Here we present a method for learning qualitative constraint trees implemented in the program QUIN (QUalitative INduction). In contrast to the usual decision trees, the leaves of *qualitative trees* contain qualitative functional constraints as used in qualitative reasoning.

The motivation for learning qualitative trees came from applications in behavioural cloning. In our previous experiments in behavioural cloning (Šuc and Bratko, 1999b; Šuc and Bratko, 1999a; Šuc and Bratko, 2000) qualitative strategies proved to be very successful. These qualitative strategies were obtained as an abstraction of quantitative control strategies induced from the human operator's execution traces. Here we investigate a natural extension of that approach, the learning of qualitative trees *directly* from numerical examples. However, the developed algorithm is not tied to learning control strategies and can be applied wherever qualitative models are preferred to the usual numerical models.

The structure of this chapter is as follows. First we introduce qualitative models and mention some related work. Section 5.2 describes our learning problem and defines the hypothesis language. In Section 5.3 we describe how qualitative constraints can be learned from a set of numerical examples. This provides the basis for the induction of qualitative trees in Section 5.4. First we give a simple ep-QUIN algorithm and improve it into the QUIN algorithm. In Section 5.5 we experimentally evaluate learning qualitative trees in artificial domains and then draw conclusions.

## 5.1 Qualitative Models

It has been shown in various studies that for some tasks qualitative models are more suitable than classical quantitative (numerical) models. These tasks include diagnosis (Bratko et al., 1989), generating explanation of the system's behaviour (Falkenhainer and Forbus, 1990) and designing novel devices from first principles (Williams, 1990).

Qualitative reasoning about processes (QPT-Qualitative Process Theory, Forbus, 1984) and qualitative simulation (QSIM, Kuipers, 1986) with Qualitative Differential Equations enable a kind of commonsense reasoning by abstracting numerical values into qualitative values and real functions into qualitative constraints.

Besides qualitative reasoning and qualitative simulation, qualitative models can also be used to guide machine learning (Clark and Matwin, 1993) and offer a space for solution optimization (Varšek et al., 1993; Šuc and Bratko, 1999b). However, building a qualitative model for a complex system requires significant knowledge and is a time-consuming process.

For this reason, many researchers are addressing the problem of automatic generation of a qualitative model. One approach is to build models from existing libraries of model fragments (Forbus, 1984). Another approach is to learn a model of a physical system from behaviours using existing knowledge of processes and mechanisms commonly found in physical systems (Doyle, 1988). Less knowledge-intensive approaches (Coiera, 1989; Bratko et al., 1991; Džeroski and Todorovski, 1993; Richards et al., 1992) use inductive learning of qualitative differential equations from a set of qualitative behaviours.

Qualitative reasoning about control includes deriving controllers by qualitative reasoning about differential equations models (Makarovič, 1991), applying qualitative simulation to the analysis of heterogeneous fuzzy controllers (Kuipers and Åström, 1994) and deriving a qualitative control rule from a qualitative differential equations model of the controlled system (Bratko, 1995). DeJong (DeJong, 1994) based his "explanation-based control" on a qualitative model of the controlled system.

## 5.2 Learning Problem Description

### 5.2.1 Qualitative Trees

We consider the usual setting of a classification learning problem, but in our case the hypothesis language involves *qualitative constraints*.

Let there be $N$ learning examples. Each example is described by $n+1$ continuous variables $X_1, \ldots, X_{n+1}$ with values $x_{i,1}, \ldots, x_{i,n+1}$, $i = 1, \ldots, N$. The variable $X_{n+1}$ is called the *class*, and others are called *attributes*. Discrete ordinal attributes can also be used and are treated similarly to continuous attributes as described in Section 5.4.4. In the context of modelling dynamic systems, an example can be a state in the system's state space. Given the learning examples, our problem is to learn a hypothesis that separates the areas of state space which share a common qualitative behaviour of the class variable. We learn such hypotheses in the form of *qualitative trees*. A qualitative tree is a binary tree with internal nodes called splits and *qualitatively constrained functions* in the leaves. The splits define a partition of the state space into areas with common qualitative behaviour of the class variable. A split consists of a split attribute and a split value. Qualitatively constrained functions (abbreviated QCFs) in leaves define qualitative constraints on the class variable, i.e give constraints on the class change in response to a change in attributes.

### 5.2.2 Qualitatively Constrained Functions

We use qualitatively constrained functions to define qualitative constraints on the class variable. A QCF constrains the qualitative change of the class variable in response to the qualitative changes of the attributes. A *qualitatively constrained function* $M^{s_1,\ldots,s_m} : \Re^m \mapsto \Re$, $s_i \in \{+,-\}$ represents an arbitrary function with $m \leq n$ continuous attributes that respect the qualitative constraints given by signs $s_i$. The qualitative constraint given by sign $s_i = +$ ($s_i = -$) requires that the function is strictly increasing (decreasing) in its dependence on the $i$-th attribute. We say that the function is *positively related* (negatively related) to the $i$-th attribute. $M^{s_1,\ldots,s_m}$ represents any function which is for all $i = 1, \ldots, m$ positively (negatively) related to the $i$-th argument if $s_i = +$ ($s_i = -$). In the case of continuously derivable functions, $s_i = +$ ($s_i = -$) requires that the $i$-th partial derivative is strictly positive (negative):

$$M^{s_1,\ldots,s_m} : \Re^m \mapsto \Re \,,\ s_i \in \{+,-\}$$
$$\frac{\partial M^{\ldots,s_i,\ldots}(\ldots,x_i,\ldots)}{\partial x_i} \begin{cases} > 0, \text{if } s_i = + \\ < 0, \text{if } s_i = - \end{cases} \tag{5.1}$$

This notation is similar to the qualitative proportionality predicates $Q_+$ and $Q_-$ as defined by Forbus (Forbus, 1984). It is also a generalization of the well known qualitative constraint $M^+(x)$, as used in (Kuipers, 1986).

Note that the qualitative constraint given by sign $s_i = +$ only states that when the $i$-th attribute increases, the QCF will also increase, *barring other changes*. It can happen that a QCF with the constraint $s_i = +$ decreases even if the $i$-th attribute increases because of a change in another attribute. One such example, i.e. the behaviour of gas pressure in a container, is given in Section 4.1.2 on page 42.

QCFs are concerned with qualitative changes and qualitative change vectors. *Qualitative change* $q_j$ is the sign of change in continuous variable $X_j$, where $q_j \in \{pos, neg, zero\}$, corresponding to *positive*, *negative* or *zero* change. A *qualitative change vector* is a vector of qualitative changes of the variables. For example, when heating the gas in the container, the observed qualitative change vector could be $(q_{Temp}{=}pos, q_{Vol}{=}zero, q_{Pres}{=}pos)$ describing the increasing temperature, steady volume and increasing pressure.

We define *QCF-prediction* $P(s_i, q_i)$ as:

$$P(s_i,q_i) = \begin{cases} \text{pos, if } (s_i{=}+ \wedge q_i{=}pos) \vee (s_i{=}- \wedge q_i{=}neg) \\ \text{neg, if } (s_i{=}+ \wedge q_i{=}neg) \vee (s_i{=}- \wedge q_i{=}pos) \\ \text{zero, otherwise} \end{cases}$$

A qualitative change vector $q = (q_1,\ldots,q_{n+1})$ is *consistent* with a given QCF $M^{s_1,\ldots,s_m}$ ($m \leq n$), if the QCF does not reject the qualitative change of the class variable, that is if either (a) class qualitative change is zero, (b) all attribute's QCF-predictions are zero, or (c) there exists an attribute whose QCF-prediction is equal to the class's qualitative change:

$$q \text{ consistent with } M^{s_1,\ldots,s_m} \text{ if either} \begin{cases} q_{n+1} = \text{zero} \\ \forall i \in \{1,\ldots,m\}\colon q_i = \text{zero} \\ \exists i \in \{1,\ldots,m\}\colon P(s_i,q_i) = q_{n+1} \end{cases}$$

This definition reflects an interesting property of the QCF called *qualitative ambiguity*. A QCF does not always uniquely predict the class's qualitative change given the qualitative changes of the attributes. Qualitative ambiguity, i.e. ambiguity in the class's qualitative change appears whenever there exist both positive and negative QCF-predictions or whenever all QCF-predictions are zero. In this case any qualitative class change is consistent with the QCF.

Note that *any* class change (not just zero class change) is consistent with a QCF when all QCF-predictions are zero. This is a weaker definition of consistency than one might expect. The reason for such definition is in learning QCFs where each possible QCF is penalized according to inconsistent qualitative change vectors: small changes in all attributes (all zero QCF-predictions) are not a strong evidence against large (nonzero) class change.

## 5.3 Learning Qualitatively Constrained Functions

When learning a QCF from a set of numeric examples, we prefer a QCF that is consistent with most of the examples, i.e. to find the *"most consistent"* QCF. For this reason we define *error-cost* $E(g)$ of a QCF $g$ (defined later in Equation 5.4 and 5.12) that penalizes with $g$ inconsistent and ambiguous qualitative change vectors at every example. The "most consistent" QCF is learned from a set of numeric examples by first forming qualitative change vectors from examples and then minimizing error-cost of a QCF over all possible QCFs. The algorithm is given in Figure 5.1 and is explained below.

*Algorithm* mostConsistentQCF($S$, $Class$, $Attributes$)
*Input:* $S$ are the training examples. $Class$ is the attribute whose qualitative change is to be predicted by the induced QCF. $Attributes$ are the attributes that may be used in the induced QCF.
*Output:* a QCF that is "most consistent" with examples $S$

1. for each pair of examples $e$ and $f$ from $S$
2. form a qualitative change vector $q_{(e,f)}$ according to Equation 5.2
3. remove qualitative change vectors with zero class change
4. $\mathcal{G}$ := set of all possible QCFs using $Attributtes$
5. $g_{best} := arg\min_{g \in \mathcal{G}} E(g)$ {E(g) is error-cost of QCF $g$}
6. **return** $g_{best}$

Figure 5.1: Algorithm mostConsistentQCF: the "most consistent" QCF is found by a simple exhaustive search algorithm that forms all possible QCFs and selects the one with the smallest error-cost. To compute error-cost $E(g)$ Equation 5.4 is used.

First, every pair of examples $e$ and $f \neq e$ is used to form a qualitative change vector with qualitative changes $q_{(e,f),j} \in \{pos, neg, zero\}, j = 1, \ldots, n+1$ defined as:

$$q_{(e,f),j} = \begin{cases} \text{pos, if } x_{f,j} > x_{e,j} + Tzero_j \\ \text{neg, if } x_{f,j} < x_{e,j} - Tzero_j \\ \text{zero, otherwise} \end{cases} \tag{5.2}$$

where $Tzero_j$ denotes a user-defined steady threshold defining negligible changes of $j$-th attribute. The default value of $Tzero_j$ is 1% of the difference between maximal and minimal value of $j$-th attribute. Typically, many pairs of examples map into the same qualitative change vector. A qualitative change vector is either consistent or not consistent with a given QCF. Note that a consistent qualitative change vector can also be ambiguous for a given QCF as discussed in Section 5.2.2. Since a qualitative change vector with zero class change is consistent with any QCF, all such qualitative vectors are removed and are not used for learning.

To learn the "most consistent" QCF we define error-cost that penalizes QCF according to the number of inconsistent and ambiguous qualitative change vectors. The error-cost is based on the minimum description length principle (Rissanen, 1978; Quinlan and Rivest, 1989). Basically it is defined as the number of bits needed to code the QCF plus the number of bits to code the inconsistent and ambiguous qualitative change vectors.

Let $\mathcal{C}_e(q)$ and $n_e(q)$ denote respectively the set and the number of all examples $f$ that form with $e$ qualitative change vector $q$:

$$\begin{aligned} \mathcal{C}_e(q) &= \{f | \forall j = 1, \ldots, n+1 : q_{(e,f),j} = q_j\} \\ n_e(q) &= |\mathcal{C}_e(q)| \end{aligned} \tag{5.3}$$

The *error-cost* $E(g)$ of a QCF $g$ that mentions $m$ out of all $n$ attributes is:

$$\begin{aligned} E(g) = & \log_2 n + m(\log_2 n + 1) + \\ & \log_2 N_{nonamb} + N_{reject}(\log_2 N_{nonamb}) + \\ & \log_2 N_{amb} + N_{amb} \end{aligned} \tag{5.4}$$

Here $N_{reject}$ denotes the number of example pairs $(e, f \neq e)$ that form a qualitative change vector that is not consistent with (is rejected by) QCF $g$ and is computed as sum of $n_e(q)$ over all examples $e$ and over all qualitative change vectors $q$ that are not consistent with $g$. Similarly, $N_{amb}$ and $N_{nonamb}$ denote the sum of $n_e(q)$ of vectors $q$ that are respectively ambiguous and not ambiguous for $g$. This error-cost is based on the following coding: we code the QCF, the indices of $N_{reject}$ inconsistent qualitative change vectors where each index requires $\log_2 N_{nonamb}$ bits since ambiguous qualitative change vectors are always consistent and one bit for each ambiguous qualitative change vector.

Given error-cost penalizes inconsistent and ambiguous qualitative change vectors formed from every pair of examples. We later define (Equation 5.12) a different error-cost and refer to here defined error-cost also as *ep-QUIN error-cost* (ep standing for every pair).

The "most consistent" QCF is found by a simple exhaustive search algorithm that forms all possible QCFs and selects the one with the smallest error-cost. This requires $2^{n+1} - 2$ computations of error-cost, where $n$ is the number of attributes. A suboptimal greedy algorithm to find the "most consistent" QCF is given in Section 5.4.3.

## 5.4 Learning Qualitative Tree

In this section we describe how a qualitative tree is learned from a set of numeric examples. The algorithm uses top-down greedy approach and is guided by error-cost of a QCF. We define two learning algorithms that use different error-costs. First we give the ep-QUIN algorithm that uses error-cost defined in Section 5.3. In Section 5.4.2 an example showing the myopia of ep-QUIN algorithm is presented. In Section 5.4.3 we give an improvement of ep-QUIN, i.e. QUIN algorithm.

### 5.4.1 Algorithm ep-QUIN

The ep-QUIN algorithm (see Figure 5.2) uses top-down greedy approach similar to the ID3 algorithm. Given the examples, ep-QUIN chooses the best split by comparing the partitions of the examples they generate: for every possible split, it splits the examples into two subsets (according to the split), finds the "most consistent" QCF in both subsets, and selects the split which minimizes the *tree error-cost* (defined below). It puts the best split in the root of the tree and recursively finds both subtrees for the corresponding example subsets, until the stopping condition is satisfied, i.e. until using the best split does not improve the tree error-cost.

*Algorithm* ep-QUIN($S$, $Class$, $Attributes$)
*Input:* $S$ are the training examples. $Class$ is the attribute whose qualitative change is to be predicted by the induced qualitative tree. $Attributes$ are attributes that may be tested in the internal nodes and used in the QCFs in leaves.
*Output:* Induced qualitative tree

```
create a Root node for the tree
g := mostConsistentQCF(S,Class,Attributes)
E(g) := error-cost of QCF g ; E_best := E(g)
for each A from Attributes do begin
    for each possible value v_i of A do begin
        split examples S into S_leq and S_grt according to test A ≤ v_i
        E_left := error-cost of mostConsistentQCF(S_leq,Class,Attributes)
        E_right := error-cost of mostConsistentQCF(S_grt,Class,Attributes)
        E_tree := tree error-cost (Equation 5.5)
        if E_tree < E_best then begin
            E_best := E_tree ; A_best := A ; v_best := v_i
        end
    end
end
if E_best < E(g) then begin
    put test A_best ≤ v_best into Root
    split examples S into S_leq and S_grt according to A_best ≤ v_best
    put subtrees obtained by ep-QUIN(S_leq, Class, Attributes)
        and ep-QUIN(S_grt, Class, Attributes) bellow Root
end else Root is a leaf with QCF g
return Root
```

Figure 5.2: Qualitative tree learning algorithm ep-QUIN.

The *tree error-cost* is computed as follows. The tree error-cost of a leaf is the error-cost $E(g)$ (Equation 5.4) of the QCF $g$ that is induced from the examples in the leaf. The tree error-cost of an internal node is the sum of error-costs of both subsets plus the cost of the split, i.e. the number of bits needed to encode the split:

$$\begin{aligned} E_{tree} &= E_{left} + E_{right} + SplitCost \\ SplitCost &= \log_2 n + \log_2(Splits_i - 1) \end{aligned} \tag{5.5}$$

Here $E_{left}$ and $E_{right}$ denote the error-costs in both subsets, $n$ is the number of variables and $Splits_i$ is the number of possible splits for the split variable, i.e. the number of different values of the variable $X_i$.

### 5.4.2 An Example

Here we give a simple example showing the myopia of ep-QUIN's error-cost. Figure 5.3 shows the learning dataset consisting of 12 learning examples with one attribute $x$ and class

$y$. The examples correspond to 3 linear functions that are increasing in its argument $x$:

$$y = \begin{cases} 2x + 7, \text{x= } 0, 1, 2, 3 \\ 2x - 16, \text{x= } 10, 11, 12, 13 \\ 2x - 39, \text{x= } 20, 21, 22, 23 \end{cases} \tag{5.6}$$

An ideal qualitative tree would divide the attribute space into 3 regions corresponding to definition areas of 3 linear functions. One such qualitative tree that is consistent with all the examples is:

$$\begin{array}{l} x \leq 3: \quad y = M^+(x) \\ x > 3 \\ \mid \quad x \leq 13: \quad y = M^+(x) \\ \mid \quad x > 13: \quad y = M^+(x) \end{array} \tag{5.7}$$

Given these examples the ep-QUIN algorithm proceeds as follows. It first finds the "most-consistent" QCF for the whole attribute space and then tries to reduce the error-cost by splitting the attribute space into subregions. There are 12 examples; therefore ep-QUIN forms $12 \times (12-1)$=132 qualitative change vectors. With $x$ as the only attribute there are only two possible QCFs: QCF $y = M^+(x)$ that is consistent with 50 (out of 132) qualitative change vectors and $y = M^-(x)$ that is consistent with 82 qualitative change vectors. Split minimizing ep-QUIN error-cost is $x \leq 1$, splitting the examples into two subsets. In the first subset ($x \leq 1$) there are only two examples forming two qualitative change vectors, both consistent with $y = M^+(x)$. In the second subset ($x > 1$) there are 10 examples forming $10 \times 9$ qualitative change vectors, 58 of them are consistent with QCF $y = M^-(x)$. The second subset is further divided resulting in the following qualitative tree:

$$\begin{array}{l} x \leq 1: \quad y = M^+(x) \\ x > 1 \\ \mid \quad x \leq 21 \\ \mid \quad \mid \quad x \leq 11: \quad y = M^-(x) \\ \mid \quad \mid \quad x > 11: \quad y = M^-(x) \\ \mid \quad x > 21: \quad y = M^+(x) \end{array} \tag{5.8}$$

This qualitative tree is clearly not optimal. The reason is that ep-QUIN does not consider the locality of qualitative changes. A human observing Figure 5.3 might notice 3 groups of near-by example points and consider the proximity of examples when estimating qualitative changes at a particular example point. In this way a qualitative change of nearby points ($x$=1 and $x$=2 for example) would be weighted more than a qualitative change of far-away points ($x$=1 and $x$=13). One might also consider the consistency of a qualitative class change for a particular vector of qualitative attributes changes. One might asses consistency of qualitative class change for a particular vector of qualitative attributes changes $q_x$ by considering 3 nearest neighbours (ignoring the class distance) of a given point that form $q_x$ with this point. Then the confidence in the positive class change ($q_y$=$pos$) for vector of qualitative attributes changes $q_x$=$pos$ at point ($x$=0,$y$=7) in Figure 5.3 would be higher than at point ($x$=2,$y$=11) since all three corresponding neighbours (with $q_x$=$pos$) of the first point give positive class change whereas the second point has one neighbour giving positive class change

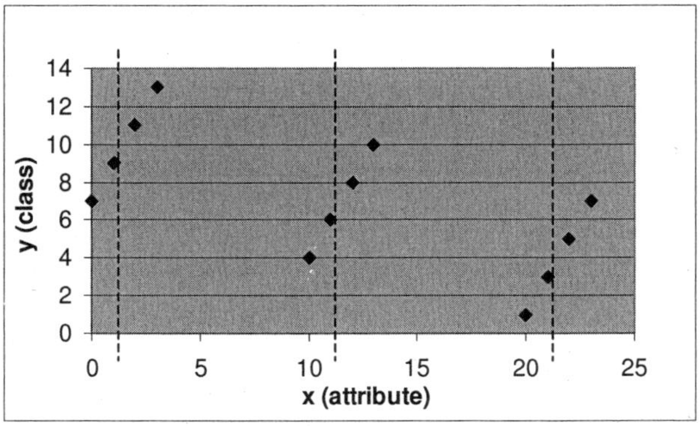


Figure 5.3: Example partition induced by ep-QUIN: the points correspond to 12 learning examples with attribute $x$ and class $y$. The vertical lines denote the example partition (see Rule 5.8) induced by ep-QUIN. Clearly this example partition is not optimal. We would prefer a partition (i.e. a qualitative tree) that divides the examples into 3 regions that correspond to definition areas of 3 linear functions from Equation 5.6 and to induce QCF $y = M^+(x)$ in each region (see Rule 5.7).

and two neighbours giving negative class change. Note that the corresponding neighbours (with $q_x$=$pos$) of the first point are the points at $x$=1,2,3, and the corresponding neighbours of the second point are the points at $x$=3,10,11.

An improvement of ep-QUIN, the QUIN algorithm follows this idea, i.e. it also considers the locality and consistency of qualitative change vectors. Note that the QUIN algorithm, from this learning dataset, induces more preferable qualitative tree (Rule 5.7) that is consistent with all the examples.

### 5.4.3 Algorithm QUIN

The QUIN learning algorithm is an improvement of the ep-QUIN algorithm that also takes the locality and consistency of qualitative change vectors into account. The algorithms differ in their error-cost and the algorithm to select the "most consistent" QCF. ep-QUIN performs exhaustive search to select the "most consistent" QCF, while QUIN uses a heuristic greedy algorithm. ep-QUIN's error-cost (Equation 5.4) uses only counts of consistent and inconsistent qualitative change vectors. QUIN's error-cost (defined later in Equation 5.12) uses heuristic confidence estimates of qualitative change vectors that also take the locality and consistency of qualitative change vectors into account.

In the following paragraphs we first define the confidence estimates of qualitative change vectors and then give QUIN's error-cost. At the end we give QUIN's greedy algorithm that finds a (suboptimal) "most consistent" QCF.

#### Estimating Qualitative Change Vectors

The QUIN learning algorithm uses heuristic confidence estimates of qualitative change vectors at example $e$. We refer to confidence estimates simply as estimates. These estimates take

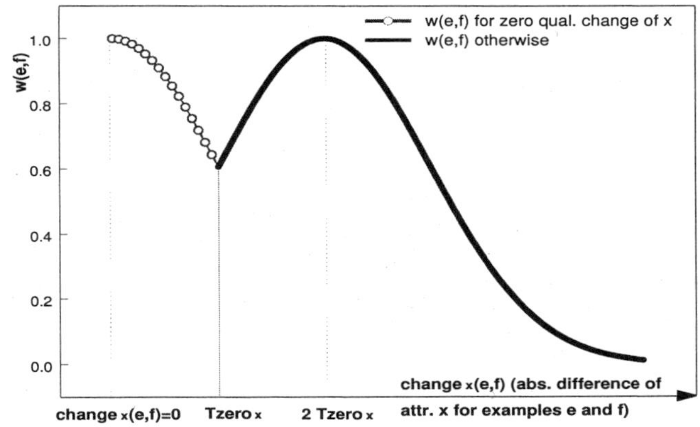


Figure 5.4: Evaluating the proximity of examples: weight $w(e, f)$ (Equation 5.10) as a function of $change_x(e, f)$, i.e. the absolute difference of examples $e$ and $f$ in the only attribute $x$. Note that $w(e, f)$ weighs attribute's value change ($change_x(e, f)$) relative to the "center point" of the qualitative change, i.e. the difference from 0 for $zero$ qualitative change and the difference from $2 \times Tzero_x$ for $pos$ or $neg$ qualitative change.

the locality and consistency of a qualitative change vector into account. First we give some necessary definitions and then describe how these estimates are computed.

We say that example $f$ is a *q-neighbour* of example $e$ if the qualitative change vector $q_{(e,f)}$ is equal to $q$ in all the qualitative attribute changes (but not necessarily in the qualitative class change). Let $\mathcal{N}_e(q)$ denote the set of all $q$-neighbours of $e$. We define the distance $d(e, f)$ (of the examples $e$ and $f$) that considers only the attributes of the two examples and $\mathcal{N}_e^k(q)$ as the set of $k$ *nearest q-neighbours of example* $e$:

$$\begin{aligned}
\mathcal{N}_e(q) &= \{f \mid \forall j = 1, \ldots, n : q_{(e,f),j} = q_j\} \\
\mathcal{N}_e^k(q) &= \text{set of } k \text{ examples } f \in \mathcal{N}_e(q) \text{ minimizing } d(e, f) \\
d(e, f) &= \sum_{j=1}^{n} \left(\frac{d_j(e, f)}{norm_j}\right)^2 \qquad (5.9) \\
d_j(e, f) &= \begin{cases} |x_{e,j} - x_{f,j}|, \text{ if } q_{(e,f),j} = \text{zero} \\ |x_{e,j} - x_{f,j}| - 2 \times Tzero_j, \text{ otherwise} \end{cases}
\end{aligned}$$

Here $norm_j$ is the normalization factor of the attribute $X_j$, computed as the difference between maximal and minimal value of $X_j$ in the learning examples. To evaluate the proximity of the examples we use Gaussian kernel weighting similar as in Section 2.4.2. The proximity of the examples $e$ and $f$ is defined as the weight $w(e, f)$ (see Figure 5.4):

$$w(e, f) = e^{\frac{-d(e,f)}{2K^2}} \qquad (5.10)$$

Here $K$ denotes Gaussian kernel that is a user-defined parameter with the default value computed as a 10% of a maximal Euclidean distance of two examples from the learning dataset. By changing $K$, more or less importance is assigned to the distance between examples and therefore more local or more global estimates are used. For example, if the user sets $K$ to

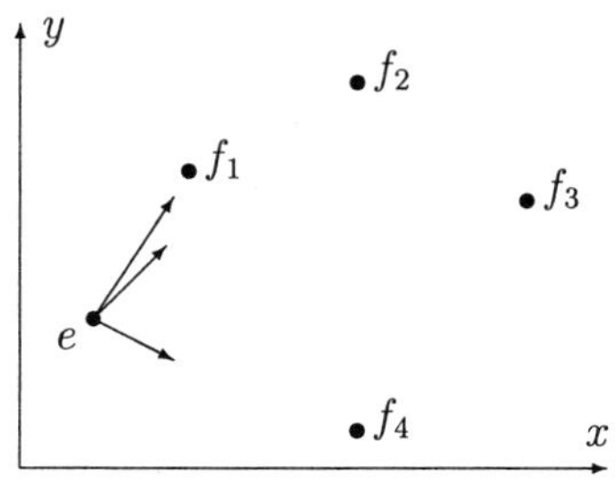


Figure 5.5: Computing estimates at the point $e$ with $k = 3$ (number of nearest $q$-neighbours used): the arrows illustrate qualitative change vectors and the lengths of the arrows indicate their weights.

1% of maximal Euclidean distance, more importance is assigned to the distance between examples, and more local estimates are used than with the default $K$ value.

The confidence estimate $w_e(q)$ of the qualitative change vector $q$ at example $e$ takes into account the proximity of the $q$-neighbours and the consistency of qualitative class change w.r.t. the $q$-neighbours. To estimate the qualitative change vector $q$ at $e$, $k$ nearest $q$-neighbours are taken into account, where $k$ is a user-defined parameter with the default value 5. The intuition in estimates $w_e(q)$ is in modelling the local probability of qualitative class change $q_{n+1}$ given the attribute's changes $q_i, i = 1, \ldots, n$:

$$\begin{aligned} \mathcal{C}_e^k(q) &= \{f \in \mathcal{N}_e^k(q) | q_{(e,f),n+1} = q_{n+1}\} \\ w_e(q) &= \frac{1}{|\mathcal{N}_e^k(q)|} \sum_{f \in \mathcal{C}_e^k(q)} w(e, f) \end{aligned} \tag{5.11}$$

Here $\mathcal{C}_e^k(q)$ denotes the set of $k$ nearest $q$-neighbours of $e$ with the qualitative class change equal as in $q$. Note that using $\left|\mathcal{N}_e^k(q)\right|$ in denominator penalizes estimate $w_e(q)$ if the nearest $q$-neighbours of $e$ differ in qualitative class change.

We illustrate this procedure by the example in Figure 5.5. In Figure 5.5 $y$ is the class variable and $x$ is the only attribute. Each pair of neighbour examples $(e, f_i)$ defines a qualitative change vector at the point $e$. The pairs $(e,f_i)$, $i$=1,2,3 have the same qualitative change vector $p$=($q_x$,$q_y$)=($pos$,$pos$), pair $(e,f_4)$ gives $r$=($q_x$,$q_y$)=($pos$,$neg$). Suppose the user set $k = 3$, so the nearest $p$-neighbours of $e$ are $\mathcal{N}_e^3(p) = \{f_1, f_2, f_4\}$. The nearest $r$-neighbours of $e$ are the same, since $p$ and $r$ differ only in qualitative class change. Now the weights $w(e, f_i)$, $i$=1,2,4 are computed. The estimates of the qualitative change vectors $p$ and $r$ at $e$ are: $w_e(p) = (w(e, f_1) + w(e, f_2))/3$ and $w_e(r) = w(e, f_4)/3$.

The idea of this estimate is similar to algorithm Relief (Kira and Rendell, 1992; Kononenko, 1994)

## QUIN's Error-Cost of a QCF

Similarly to ep-QUIN's error-cost, QUIN's error-cost is based on the minimum description length principle. It estimates the number of bits needed to code a QCF plus the number of bits to code the inconsistent and ambiguous qualitative change vectors, where the change vectors $q$ are weighted according to their confidence estimates $w_e(q)$ (eq. 5.11). Because of this weighting QUIN's error-cost is heuristic and does not correspond to a particular coding of examples.

The error-cost $E(g)$ of a QCF $g$ that mentions $m$ out of all $n$ attributes is:

$$\begin{aligned} E(g) = &\log_2 n + m(\log_2 n + 1) + \\ &\log_2 W_{nonamb} + W_{reject}(\log_2 W_{nonamb}) + \\ &\log_2 W_{amb} + W_{amb} \end{aligned} \tag{5.12}$$

Here $W_{reject}$ denotes the sum (over all examples $e$) of the estimates $w_e(q)$ of the qualitative change vectors $q$ that are not consistent with $g$. Similarly $W_{amb}$ and $W_{nonamb}$ denote the sum of the estimates $w_e(q)$ of vectors $q$ that are respectively ambiguous and not ambiguous for $g$.

### Greedy QCF Learning Algorithm

ep-QUIN uses an exhaustive search algorithm to find the "most consistent" QCF (Figure 5.1). Selecting a QCF with the smallest error-cost among all possible QCFs requires the number of error-cost computations that is exponential in the number of attributes. QUIN uses a heuristic greedy algorithm that requires the number of error-cost computations that is quadratic in the number of attributes, but finds a possibly suboptimal "most consistent" QCF.

Let $h$ denote a *single-attribute QCF* (that uses only one attribute), i.e. $h{=}M^{s_k}(a_k)$, and let $R(g,h)$ denote the refinement of QCF $g = M^{s_1,..,s_m}(a_1,..,a_m)$ with $h$: $R(g,h) = M^{s_1,...,s_m,s_k}(a_1,\ldots,a_m,a_k)$.

The greedy QCF learning algorithm is as follows. At the first step, QCF $g_{best}$ with the minimal error-cost is selected among all single-attribute QCFs. At the second step, QCF $h$ with the minimal error-cost of $R(g_{best},h)$ is selected among all single-attribute QCFs. If the error-cost of the $R(g_{best},h)$ is lower than the error-cost of $g_{best}$ and $h$ is consistent with at least two thirds (a heuristic, commented below) of qualitative change vectors that are not consistent with $g_{best}$, then $g_{best}$ is set to $R(g_{best},h)$ and the second step is repeated. Otherwise $g_{best}$ is returned as the "most consistent" QCF.

In the above algorithm, a stopping condition for refinements is also a heuristic "two thirds", requiring that a refinement of $g_{best}$ is consistent with at least two thirds of qualitative change vectors not consistent with $g_{best}$. The idea is to prevent refinements that improve the error-cost mainly by increasing ambiguity. We decided for "two thirds" since there is always a refinement of $g_{best}$ (using an arbitrary attribute) that is consistent with at least one half of qualitative change vectors that are not consistent with $g_{best}$.

## 5.4.4 Notes on Learning Algorithms

QCFs in the leaves of a qualitative tree give constraints on continuous attributes. Sometimes it is also useful to constrain the set of attributes that can be used in QCFs. The attributes that are marked as *discrete attributes* are used just for splits in the internal nodes and are not used in QCFs. Normally these are discrete ordinal attributes or expert-selected continuous attributes that are not suitable for qualitative constraints in leaves.

Since QUIN uses a greedy algorithm to select the most consistent QCF it is more time efficient than ep-QUIN. We compared the time efficiency on a domain with 200 examples using Pentium II, 400 MHz computer. When 3 attributes were used, QUIN required 38 seconds and ep-QUIN 58 seconds. When 6 attributes were used, QUIN required 3 minutes and ep-QUIN required 17 minutes. Such time complexity is mainly the result of the time-expensive algorithms for learning of the "most consistent" QCF. These algorithms have quadratic (for QUIN) or exponential (for ep-QUIN) dependency of the number of QCF error-cost computations on the number of the attributes. Computing the error-cost $E(g)$ of QCF $g$ requires to determine the consistency of every qualitative change vector with QCF $g$.

## 5.5 Experimental Evaluation of QUIN Algorithm

Here we experimentally evaluate learning qualitative trees from numerical examples on a set of artificial domains. First we define *qualitative tree performance* as a measure of qualitative tree prediction error. Then we describe the details of our experimental evaluation and give results.

### 5.5.1 Qualitative Tree Performance

Since the usual error measures, such as accuracy or mean squared error, are not suitable for qualitative models, we define the measure of qualitative tree prediction error. We define *qualitative tree performance* (abbreviated as QP) as a pair of two percentages:

- *Qualitative consistency* is the percentage of qualitative change vectors that are consistent with a qualitative tree. The examples are partitioned according to the splits of the tree and for each example partition the qualitative change vectors are formed. Consistency is the percentage of qualitative change vectors that are consistent with QCFs in the corresponding leaves.

- *Qualitative ambiguity* is the percentage of qualitative change vectors that are ambiguous for a qualitative tree, i.e. the percentage of qualitative change vectors which give ambiguous QCF-prediction in the corresponding leaves. Ambiguity is computed similarly to consistency, by counting the number of ambiguous qualitative change vectors and the number of all qualitative change vectors in each example partition.

When giving the qualitative tree performance, we use the notation QP=$c/a$, where $c$ denote the consistency and $a$ ambiguity of a qualitative tree on the example set. We prefer a qualitative tree with high consistency and low ambiguity. However, low ambiguity is not possible when many attributes are used in the same QCFs.

Ambiguity is used in a qualitative tree performance since consistency of the qualitative tree does not give full information about the qualitative tree prediction strength. Consider for example a dataset with two attributes $x_1$ and $x_2$ that are pairwise equal for all the examples. Then attributes $x_1$ and $x_2$ have the same qualitative change in all qualitative change vectors. QCF $M^{+,-}(x_1, x_2)$ is consistent with all the examples, but is not useful in predicting qualitative class change since it is also ambiguous for any qualitative change vector from the dataset. The qualitative tree performance would in this case be QP=100%/100%.

Note that ep-QUIN minimizes the error-cost that is basically a tradeoff between consistency and ambiguity. QUIN minimizes error-cost that use weighted qualitative change vectors and therefore uses different notion of consistency and ambiguity. Therefore, the ep-QUIN algorithm is given an advantage when comparing QUIN and ep-QUIN by here defined qualitative tree performance.

Table 5.1: The description of the test domains: attributes $x, y \in [-20, 20]$ and $u, v \in [-1, 1]$ are uniformly distributed. The third column gives qualitative performance of the optimal (i.e. most consistent) qualitative tree.

| Domain | Class variable | Opt. QP, leaves |
|---|---|---|
| $Sin$ | $c = \sin(\pi \times \frac{x}{10})$ | 100/6, 5 |
| $SinLn$ | $c = \frac{x}{10} + sign(x) \times \sin(\pi \times \frac{x}{10}))$ | 100/4, 6 |
| $Poli$ | $c = \ln(10^4 + \|(x+16)(x+5)(x-5)(x-16)\|)$ | 100/10, 8 |
| $Signs$ | $c = \begin{cases} sign(u+0.5)(x-10)^2, \text{if } v \geq 0 \\ sign(u-0.5)(y+10)^2, \text{otherwise} \end{cases}$ | 100/2, 8 |
| $QuadA$ | $c = x^2 - y^2$ | 100/44, 4 |
| $QuadB$ | $c = (x-5)^2 - (y-10)^2$ | 100/44, 4 |
| $SQuadB$ | $c = sign(u)((x-5)^2 - (y-10)^2)$ | 100/38, 8 |
| $YSinX$ | $c = y\sin(\pi \times \frac{x}{10})$ | 100/39, 16 |

### 5.5.2 Experimental Details

The learning algorithms were evaluated on a set of artificial domains with uniformly distributed attributes $x, y \in [-20, 20]$ and $u, v \in [-1, 1]$. The class variables $c$, i.e. the variables to be predicted by a qualitative trees were computed as given in Table 5.1. For learning we used just the relevant attributes plus 2 uniformly distributed (noise) attributes; in domain $Sin$, for example, we used the relevant attribute $x$ and two uniformly distributed irrelevant attributes. The table also gives the number of leaves and qualitative performance (measured on the test set) of the optimal qualitative trees, i.e. the smallest trees with 100% consistency. An example of such an optimal qualitative tree for the domain $QuadB$ is:

$$\begin{array}{l} x \leq 5 \\ \mid \;\; y \leq 10 : c = M^{-,+}(x, y) \\ \mid \;\; y > 10 : c = M^{-,-}(x, y) \\ x > 5 \\ \mid \;\; y \leq 10 : c = M^{+,+}(x, y) \\ \mid \;\; y > 10 : c = M^{+,-}(x, y) \end{array}$$

Although optimal trees are quite simple, we sometimes might prefer a smaller tree or a tree with lower ambiguity. An example is the domain $YSinX$ (see Figure 5.6), where optimal tree has 16 leaves and ambiguity 39%. A smaller tree with 10 leaves has considerably lower ambiguity (10%) and consistency 96%.

Note also that the ambiguity of the optimal trees is not zero. This is either because QCFs in leaves use more than one attribute or because of zero change of all the attributes that appear in a QCF. Zero changes are defined by steady threshold parameter $Tzero_j$. We used the default value, that is 1% of the difference between maximal and minimal value of $j$-th attribute. Other parameters of the QUIN algorithm were also set to its default values: Gaussian kernel $K$ is 10% of maximal Euclidean distance of the two examples from the learning dataset and $k$ (the number of nearest $q$-neighbours used) is 5.

ep-QUIN and QUIN produce trees of different sizes; QUIN usually produces smaller trees. To evaluate the qualitative performance, we limited the size of constructed tree to 10

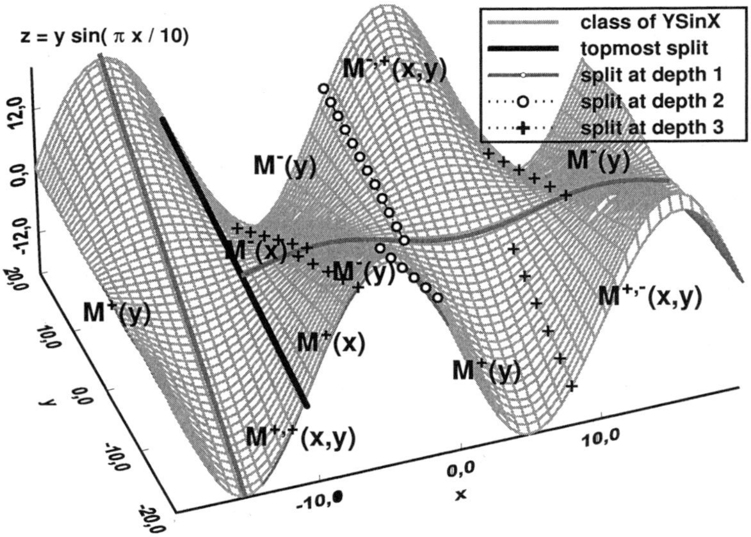


Figure 5.6: The attribute space partition induced by QUIN in domain $YSinX$: the surface is the class variable, i.e. $y \sin(\pi \times \frac{x}{10})$ and the lines are the induced splits that correspond to a qualitative tree with 10 leaves and qualitative performance 88%/28%.

leaves by pruning the nodes with the smallest number of examples because we prefer small qualitative trees.

### 5.5.3 Experimental Results

To compare the qualitative performance of the trees produced by the two algorithms, we used learning sets consisting of 200 examples and test sets of 500 examples. All the results are averages of 10 runs: 10 sets of examples were used for learning and the induced tree was tested on a separate test set consisting of 500 examples.

Figure 5.7 gives qualitative performance for qualitative trees produced by ep-QUIN and QUIN algorithms. Qualitative performance for each domain is given on $x$-axis: gray color corresponds to ep-QUIN and black color to QUIN algorithm, diamonds denote qualitative consistency and triangles ambiguity. QUIN has better qualitative performance, i.e. higher qualitative consistency and lower ambiguity on $SinLn$, $Signs$, $QuadA$ and $QuadB$. On $SQuadB$ the differences are not so obvious, but still QUIN might be a bit better. On $Poli$, QUIN is better since its ambiguity is considerably smaller. On $Sin$ QUIN usually finds optimal tree whereas ep-QUIN is slightly worse and produces bigger trees. On $YSinX$ ep-QUIN produces trees with higher consistency, but also much higher ambiguity; the QUIN's qualitative performance is closer to already mentioned preferred tree with 10 leaves that has consistency 96% and ambiguity 10%.

On most of the domains, QUIN usually induced a "near-optimal" tree, i.e. small tree with consistency over 96%. The exception is the domain $YSinX$, where ep-QUIN and QUIN fail. Figure 5.6 gives the attribute space partition and the induced QCFs that correspond to a qualitative tree induced by QUIN. This qualitative tree with 10 leaves and qualitative

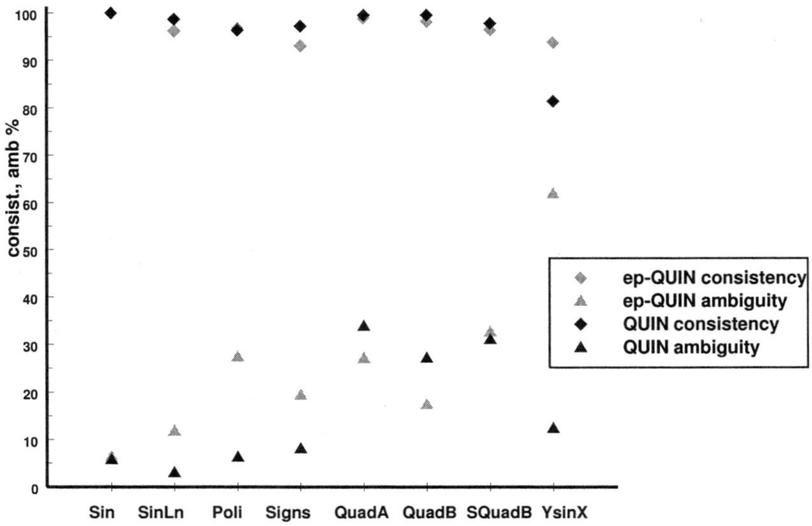


Figure 5.7: Qualitative performance of ep-QUIN's and QUIN's trees: for each domain on $x$-axis, qualitative performance for learning a set of 200 examples is given. Gray color corresponds to ep-QUIN and black to the QUIN algorithm.

performance 88%/28% is not close to the optimal qualitative tree. We give an example of a "near-optimal" qualitative tree with qualitative performance 98%/2% induced in the domain $Signs$:

$v \leq 0.02$
| $y \leq -9.89$
| | $u \leq 0.49 : c = M^{+}(y)$
| | $u > 0.49 : c = M^{-}(y)$
| $y > -9.89$
| | $u \leq 0.49 : c = M^{-}(y)$
| | $u > 0.49 : c = M^{+}(y)$
$v > 0.02$
| $x \leq 8.26$ {opt. split is x $\leq$ 10 }
| | $u \leq -0.51 : c = M^{+}(x)$
| | $u > -0.51 : c = M^{-}(x)$
| $x > 8.26$ {opt. split is x $>$ 10 }
| | $u \leq -0.49 : c = M^{-}(x)$
| | $u > -0.49 : c = M^{+}(x)$

We also observed how qualitative performance changes with the increasing learning set size. Qualitative trees induced from $20, 40, \ldots, 180$ learning examples were tested on a test set of 500 examples. Figure 5.8 gives two typical learning curves of qualitative trees induced by QUIN and ep-QUIN. The learning curve for domain $Sin$ is on the left graph and the

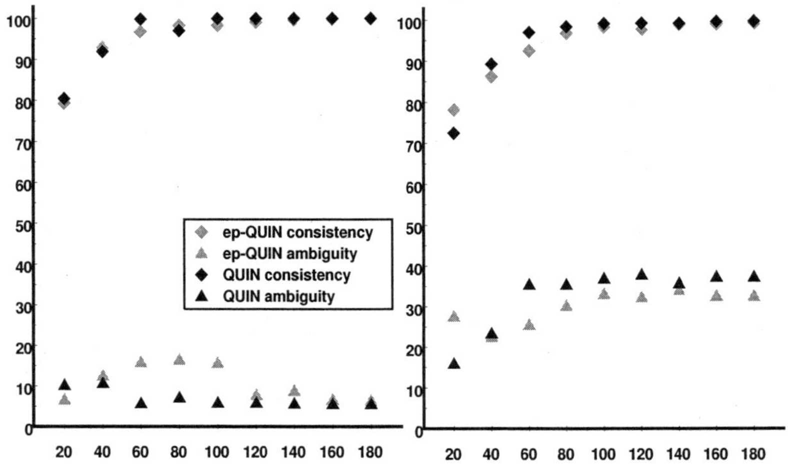


Figure 5.8: Learning curves of ep-QUIN and QUIN trees on domains $Sin$ (left graph) and $QuadA$ (right graph): on $x$-axis is the number of learning examples and on $y$-axis is qualitative performance of the induced trees.

learning curve for domain $QuadA$ is on the right graph. All results are averages of 10 runs.

When we compared the trees induced by ep-QUIN and QUIN, we observed that QUIN trees are usually smaller. An example is the domain $Sin$ where QUIN usually induces optimal tree with 5 leaves, whereas ep-QUIN induces larger tree with high consistency. Figure 5.9 gives ep-QUIN's QUIN's tree error-cost (Equation 5.5) for the only relevant attribute $x$ in the domain $Sin$. Both algorithms use tree-error cost to choose the best split: the split that minimizes tree-error cost is put into the root of the qualitative tree. The minima of QUIN tree error-cost ($x$=5,-5,-15, 15) correspond to splits dividing $sin(\pi x/10)$ into monotonic areas, whereas minimum of ep-QUIN tree error-cost ($x$=0) is not a good split since it splits a monotonic area $x \in [-5, 5]$. QUIN puts the split $x \leq 5$ in the root of the qualitative tree, whereas ep-QUIN chooses split $x \leq 0$ as the topmost split.

Since the QUIN tree error-cost also considers the locality of qualitative change vectors, it is a better criterion of split goodness. For this reason, QUIN often produces smaller trees with higher consistency.

### 5.5.4 Robustness against Noise in the Class

We also experimented with noise in the class: normally distributed noise with mean 0 and different standard deviation was added to the class variable. We used learning sets of 100 examples and test sets of 500 examples. The results are averages of 10 runs. Figure 5.10 gives two typical noise curves of qualitative trees induced by QUIN and ep-QUIN. The noise curve for domain $Sin$ is on the left graph and the noise curve for domain $QuadA$ is on the right graph. Note that the noise is very high: for $Sin$ the maximal standard deviation 1.4 corresponds to 70% of the class range (2) and for $QuadA$ the maximal standard deviation 180 corresponds to 22.5% of the class range (800). In the domain $Sin$ (left graph), QUIN has higher consistency than ep-QUIN when the noise is low. When the standard deviation of noise is higher than 0.5, ep-QUIN's consistency is higher but at the cost of very high

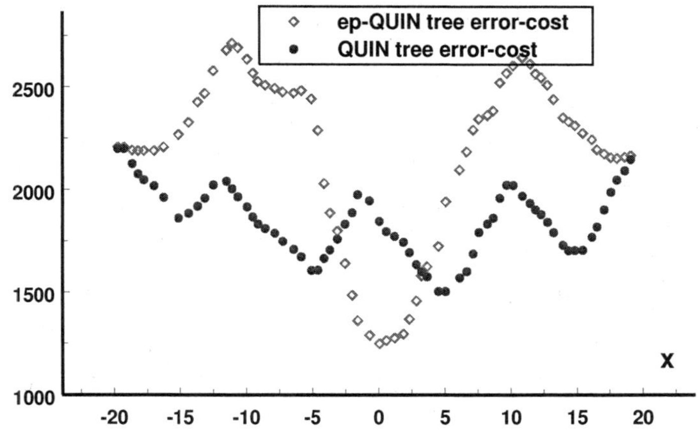


Figure 5.9: ep-QUIN's and QUIN's criterion of split goodness (i.e. tree error-cost) for attribute $x$ in domain $Sin$. Note that $x$ values with minima of QUIN's error cost ($x$=5,-5,-15, 15) correspond to points dividing $sin(\pi x/10)$ into monotonic areas.

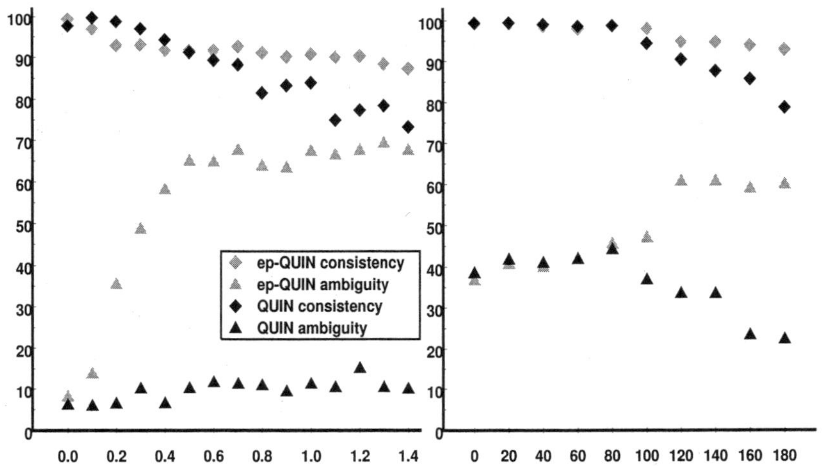


Figure 5.10: Noise curves of ep-QUIN's and QUIN's trees on domains $Sin$ (left graph) and $QuadA$ (right graph): the standard deviation of normally distributed noise that was added to the class of the learning examples is on $x$-axis and the qualitative performance of the induced trees is on $y$-axis.

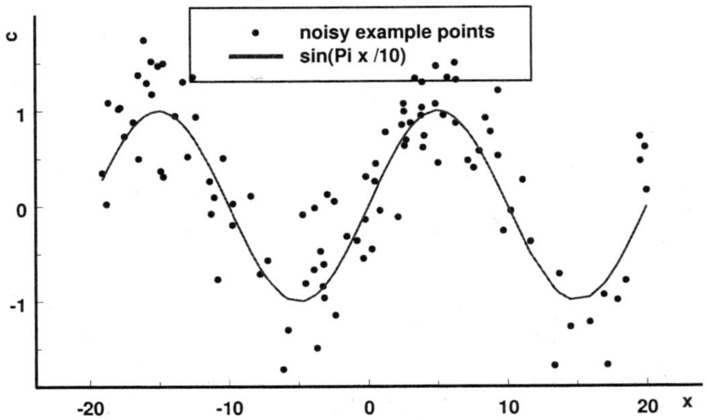


Figure 5.11: Noisy domain $Sin$: normally distributed noise with standard deviation 0.5 is added to the class variable. Qualitative trees induced by QUIN have qualitative consistency over 90%.

ambiguity. QUIN's ambiguity is small even with very high noise. In the domain $QuadA$ (right graph) ep-QUIN's ambiguity increases, whereas QUIN's ambiguity decreases when the noise is high.

The explanation for this behaviour in QUIN and ep-QUIN performance is as follows. When the noise is high, there are many inconsistent qualitative change vectors. These have small estimates $w_e(q)$ and QUIN does not try to cover them. ep-QUIN, on the other hand, does not consider the consistency of qualitative change vectors and covers inconsistent qualitative change vectors with ambiguous QCF resulting in increased ambiguity.

These results suggest that QUIN is very robust against noise in the class variable. Figure 5.11 shows the learning examples where the noise in the class variable has standard deviation 0.5. Still, QUIN induces trees with average qualitative performance 91.2%/10.2%.

## 5.6 Conclusions

We developed ep-QUIN and QUIN algorithms for learning qualitative trees from numerical examples. QUIN, a heuristic improvement of ep-QUIN, considers the locality and consistency of qualitative change vectors and handles noisy examples. We compared the two algorithms on a set of artificial domains. QUIN usually produced preferable qualitative trees and is more time efficient.

# Chapter 6

# QUIN in Skill Reconstruction

In Chapter 4 we derived a qualitative control strategy as an abstraction of the induced operator's strategy. Experiments in the crane and acrobot domains demonstrated the potentials of such strategies and are the motivation to induce qualitative strategies *directly* from execution traces. For this purpose we use the program QUIN, described in the previous chapter.

First we give some notes on using QUIN to induce the operators's qualitative strategy and then present experiments in the domains of bicycle riding and controlling a crane. In Section 6.4 we discus some points of interest.

## 6.1 Using QUIN to Induce a Qualitative Strategy

QUIN is a program that induces a qualitative tree from numerical examples. In the case of the human control skill reconstruction, the learning examples are (one or more) execution traces and the induced qualitative tree is a qualitative model of the operator's control skill. We also call it a qualitative strategy.

According to our cloning problem decomposition (Section 3.1) we use QUIN to induce a generalized operator's trajectory from an operator's trace. A chosen state variable is the class and the others are the attributes. It would also be possible to induce constraints on the operator's control action. We prefer a generalized trajectory since a strategy in the form of a generalized trajectory is often easier to interpret and because learning a generalized trajectory yields more successful and robust controllers. The robustness and success of controllers is not an issue when learning a qualitative strategy, but become important if we use the induced qualitative strategy as a space for controller optimization as explained in Section 4.1.3.

## 6.2 QUIN in Bicycle Riding

In this section we apply QUIN to the control task that requires maintaining balance on the bike and driving it to the goal. A detailed description of the domain is given in Section 2.3.3.

We used execution traces to reconstruct the operator's skill to control the bike simulator. The operators found maintaining the balance on the bike simulator a very hard task that requires a precise adjustments of the front wheel direction. The requirement to approach the goal makes the task even harder since balancing and goal-aiming must be performed simultaneously.

First we give the details of using QUIN to induce operator's qualitative strategy from execution traces. Then we present an example induced from an operator's trace and show that it gives a good insight into the operator's sub-cognitive control skill. In Section 6.2.2 we show that qualitative strategies are also very general and successful when transformed into quantitative strategies. In Section 6.2.3 we test the robustness of the resulting controllers.

### 6.2.1 Learning Qualitative Strategy

#### Experimental Details

We used traces where operators controlled the front wheel angle $\theta$ by applying torque control to the handlebars. The state of the bike is described by six variables: the tilt $\omega$ angle of the bicycle from vertical and its velocity $\dot{\omega}$, the angle $\theta$ between the front wheel direction and the bicycle direction due to the deflection of the handlebars and its velocity $\dot{\theta}$, the distance $D_g$ from the goal and the angle $\Psi_g$ of the bike's frame relative to the goal. The speed of the bike is assumed to be constant.

QUIN was used to induce a qualitative tree from a successful operator's trace. Traces were sampled with $dt = 0.1$ seconds. We decided to induce the generalized trajectory of the desired front wheel angle $\theta_{des}$. Therefore $\theta_{des}$ was the class, and other state variables were the attributes. $\theta_{des}$ was computed as $\theta$ in the next state from the execution trace: $\theta_{des}(t) = \theta(t + dt)$.

Note that it would also be possible to learn the generalized trajectory of the desired torque control or $\dot{\theta}_{des}$, but the $\theta_{des}$ trajectory is (according to the operator's description of his strategy) closer to human reasoning about riding a bike and therefore easier to interpret.

When experimenting with different traces, an interesting observation was that similar trees were learned from the successful traces of the same operator. Also, most of the trees induced from a successful trace have the "default" QCF $\theta_{des} = M^{+,+,-}(\omega, \dot{\omega}, \Psi_g)$ (explained below) in the root of the tree.

#### An Example of the Induced Strategy

QUIN induced the following qualitative tree from one of the most successful execution traces:

$$
\begin{array}{l}
\Psi_g < 0.015 \\
\mid\ \ \Psi_g < -0.027 \\
\mid\ \ \mid\ \ \theta_{des} = M^{+,+,-}(\omega, \dot{\omega}, \Psi_g) \\
\mid\ \ \Psi_g > -0.027 \\
\mid\ \ \mid\ \ \theta_{des} = M^{+,+}(\omega, \dot{\omega}) \\
\Psi_g > 0.015 \\
\mid\ \ \theta_{des} = M^{+,+,-}(\omega, \dot{\omega}, \Psi_g)
\end{array}
$$

By introducing the qualitative term $NearZero(\Psi_g) \equiv (-0.027 < \Psi_g < 0.015)$ this tree can be written as the qualitative rule:

$$
\begin{array}{l}
\text{if } NearZero(\Psi_g) \text{ then} \\
\quad \theta_{des} = M^{+,+}(\omega, \dot{\omega}) \\
\text{else } \theta_{des} = M^{+,+,-}(\omega, \dot{\omega}, \Psi_g)
\end{array}
\tag{6.1}
$$

Although this rule is very simple, it provides a good insight into the operator's control strategy. The operator uses the simpler qualitative rule when the bike is oriented towards the goal (when $NearZero(\Psi_g)$ is true) and the more complex rule when the bike is not oriented towards the goal and has to turn towards the goal.

When the bike is oriented towards the goal, the operator just maintains the balance by the rule $\theta_{des} = M^{+,+}(\omega, \dot{\omega})$. The rule is very intuitive and tells that the front wheel angle $\theta$ should be bigger at bigger tilt angle $\omega$. Bigger $\omega$ requires bigger centrifugal force $F_c$, which acts from the center of the turn and prevents the fall of the bike towards the center of the turn. Bigger $F_c$ is achieved by bigger $\theta$ as the radius of the turn is smaller. The rule also tells to increase (decrease) $\theta$ when the tilt velocity $\dot{\omega}$ is increasing (decreasing). Therefore, if the bike starts falling over, the rider should turn the front wheel in the direction of the fall, which increases the restoring centrifugal force. Since increasing $|\theta|$ causes increasing $|F_c|$, the change of $\theta$ causes the opposite change of $\dot{\omega}$ and the rule aims at bounding the magnitude of $\dot{\omega}$.

More surprising and possibly counter-intuitive is the rule $\theta_{des}$=$M^{+,+,-}(\omega, \dot{\omega}, \Psi_g)$. This rule balances the bike similar to the previous rule and also turns it towards the goal by turning the front wheel in the *opposite direction, away from the goal.* One might naturally expect that turning to the goal is achieved by turning the front wheel in the direction of the goal by the rule $\theta_{des} = M^{+}(\Psi_g)$. But this gives exactly the opposite result. Suppose the bike drives straight with $\omega = \dot{\omega} = \theta = \dot{\theta} = 0, \Psi_g = -\pi/2$ and then suddenly increases $\theta$ towards the goal. This decreases $\dot{\omega}$ and $\omega$. To prevent the fall, $\theta$ must be decreased away from the goal and once the balance is restored, $\omega$ is negative. In this way, negative $\theta$ maintains the balance and the bike is turning away from the goal. To really achieve a turn towards the goal, we need to change the tilt $\omega$ (not $\theta$) towards the goal by first decreasing $\theta$ away from the goal. Only after that, $\theta_{des}$ increases since $\omega$ has increased. That is exactly what happens if the induced rule is used.

### 6.2.2 Transforming Qualitative Strategy into Quantitative Strategies

The induced qualitative strategy given by Rule 6.1 defines a set of quantitative strategies with free qualitative term $NearZero$ and qualitatively constrained functions $M^{+,+}(\omega, \dot{\omega})$ and $M^{+,+,-}(\omega, \dot{\omega}, \Psi_g)$. By concretising these parameters into a real valued constraint and real-valued functions, different quantitative strategies are obtained. We can also use some domain knowledge that imposes additional constraints on the parameters. This approach was used in the experiments in the crane and acrobot domains in Chapter 4. The resulting strategies perform the control task in a way that is qualitatively equivalent to the original operator's control style. To achieve the control style that is also quantitatively similar to the operator's, information from the operator's trace can be used when concretising the qualitative parameters. The other reason to use the operator's trace when transforming a qualitative strategy into quantitative strategies is to minimize the use of the domain knowledge.

We use QCF $M^{+,+,-}(\omega, \dot{\omega}, \Psi_g)$ to illustrate how a bike domain knowledge can be combined with a qualitative strategy. This QCF is in a leaf of the induced qualitative tree and is also the default QCF in the root of the tree. When Rule 6.1 was used instead of $M^{+,+,-}(\omega, \dot{\omega}, \Psi_g)$, the results were similar.

QCF $M^{+,+,-}(\omega, \dot{\omega}, \Psi_g)$ is transformed into real-valued functions $f$ as follows. The maximal and minimal values from the operator's trace are used to set the definition area of $f$. These values are (approximately): $\omega_{max}$=$\pi/36$, $\dot{\omega}_{max}$=0.12, $\theta_{max}$=$\pi/6$, $\Psi_{g_{max}}$=$\pi/2$. Then $f$

is a mapping:

$$f : [-\omega_{max}, \omega_{max}] \times [-\dot{\omega}_{max}, \dot{\omega}_{max}] \times [-\Psi_{max}, \Psi_{max}] \mapsto [\theta_{min}, \theta_{max}]$$

that satisfies qualitative constraints of $M^{+,+,-}(\omega, \dot{\omega}, \Psi_g)$. Three simple facts about the bike control task define additional constraints on $f$:

1. *maximal front wheel deflection*: maximal $|\theta_{des}|$ is between (depending on the operator) $\pi/10$ and $\pi/3$ ($\pi/10 \leq |\theta_{min}|, |\theta_{max}| \leq \pi/3$).
2. *drive straight if bike is aiming at the goal*: $f(0, 0, 0) = 0$.
3. *balancing is more important than aiming at the goal*: when $\omega$ and $\dot{\omega}$ both change from their minimal to their maximal values, $\theta_{des}$ changes $c > 1$ times of the $\theta_{des}$ change when $\Psi_g$ changes from its minimal to its maximal value (for arbitrary fixed $\omega$ and $\dot{\omega}$). Considering the constraints of the QCF $M^{+,+,-}(\omega, \dot{\omega}, \Psi_g)$ this gives:

$$\begin{aligned} \Delta_\theta &= f(\omega_{max}, \dot{\omega}_{max}, 0) - f(-\omega_{max}, -\dot{\omega}_{max}, 0)/c \\ &= |f(0, 0, \Psi_{max}) - f(0, 0, -\Psi_{max})| \\ &= |f(\omega_{max}, \dot{\omega}_{max}, \Psi_{max}) - f(\omega_{max}, \dot{\omega}_{max}, -\Psi_{max})| \\ &= |f(-\omega_{max}, -\dot{\omega}_{max}, \Psi_{max}) - f(-\omega_{max}, -\dot{\omega}_{max}, -\Psi_{max})| \\ f(-\omega_{max}&, -\dot{\omega}_{max}, \Psi_{max}) = \theta_{max} \\ f(\omega_{max}&, \dot{\omega}_{max}, -\Psi_{max}) = \theta_{min} \\ \theta_{max} &- \theta_{min} = (c+1)\,\Delta_\theta \end{aligned} \tag{6.2}$$

Note that at least some of this domain knowledge could also be obtained from the operator's trace.

We used the procedure described in Section 4.1.3 to generate at random above described functions $f$. Every generated function defines a quantitative strategy consistent with the induced qualitative strategy and the constraints imposed by the three simple facts of the bike control task. To generate the functions $f$ we used the grid $C = 8 \times 8 \times 8$, $c = 2, 3, 4, 5$ and random $|\theta_{min}|$, $|\theta_{max}|$ from $[\pi/10, \pi/3]$.

To control the system, the torque control was computed by a conventional PD controller (Franklin et al., 1986):

$$\begin{aligned} \tau &= P \times (\theta_{des} - \theta) - D \times \dot{\theta} \\ \theta_{des} &= f(\omega, \dot{\omega}, \Psi_g) \end{aligned}$$

We used $P = 50$ and $D = 5$ exactly the same as in (Brodie and DeJong, 1999). We could, of course, use the same approach as in our experiments in the crane and acrobot domain and use a learned local model of the system's dynamics to compute the torque aiming to achieve $\theta_{des}$. We used the same approach as in the previous work in this domain to show that using the generalized trajectory is not tied to a particular kind of the system's dynamics model. Since the induced strategy is goal-directed, it is robust also with respect to the error in the system dynamics model.

Out of 400 randomly generated quantitative strategies, 237 (59.2 %) completed the task with the mean time 42.1 sec. and best time 35.7 sec. Many of the quantitative strategies are

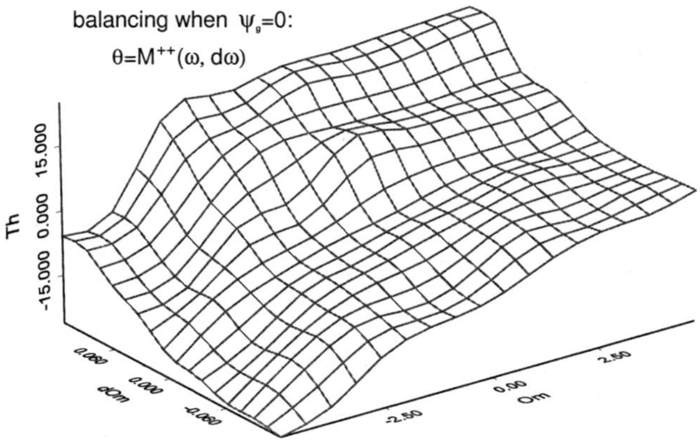


Figure 6.1: Balancing strategy of the clone in Figure 6.2: the $Th$-axis is the desired front wheel angle $\theta_{des} = f(\omega, \dot{\omega}, \Psi_g)$ at $\Psi_g$=0. The balancing strategy satisfies the induced QCF $\theta_{des} = M^{+,+}(\omega, \dot{\omega})$ and was generated at random.

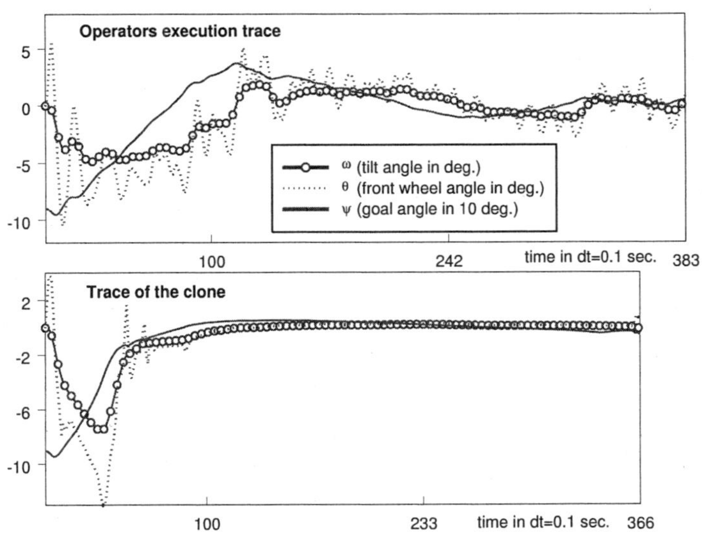


Figure 6.2: Operator's trace and its clone: the operator completes the task in 38 sec., while the clone needs approximately 36 sec. They both use the same qualitative strategy, but the clone makes the initial turn faster and is more precise in aiming at the goal afterwards.

faster than any human operator's trace. The remaining 163 strategies failed to maintain a balance or were not accurate enough in aiming at the goal. An execution trace of one of the clones and the operator's trace are given in Figure 6.2. In Figure 6.1 the clone's balancing strategy is illustrated by the generated $f(\omega, \dot{\omega}, \Psi_g)$ at $\Psi_g$=0.

### 6.2.3 Test of Robustness

Here we test the controllers from the previous section for their robustness. We performed two sets of experiments with:

1. *change in the start state:* the starting distance of the goal $D_g$ was modified by 20%, that is from $D_g$=100 in the original control task to $D_g$=80 and $D_g$=120.

2. *random displacement of the bicyclist from the center of mass:* at each time step a random $d_{noise} \in [-0.002, 0.002]$ was added to the current displacement from the mass center. If the absolute value of the displacement exceeded $d_{max}$=0.02 m, it was set to $d_{max}$. Note that in all previous experiments displacement from the mass center was zero.

As in the previous section, we randomly generated 100 quantitative strategies, corresponding to 100 controllers. Every controller was tested with both changes in the start state and with random displacement of center of mass. With $D_g$=80, out of 100 controllers 54 were successful. 60 controllers were successful with $D_g$=120. Comparing this with 59.2% controllers that are successful for the original task, we can conclude that a change in the start state does not effect the controller much.

When using random displacement from mass center 26 out of 100 controllers were successful. Considering the fact that random displacement makes a control task considerably harder the results are very good.

Some controllers were also tested with considerably different start states, for example with $D_g$=200 and changed starting orientation of the bike. Most of the controllers that were successful with the original task were also successful with altered tasks.

## 6.3 QUIN in Crane Domain

In this section QUIN is applied to the reconstruction of the operator's crane control skill. This domain is described in Section 2.3.1. In the previous study (Urbančič, 1994) that also included experiments with human learning of the crane control, remarkable individual differences were observed in the speed of control and the characteristics of the strategies operators used. Some operators tended towards fast and less reliable operation, others were more conservative and slower in order to avoid large rope oscillations.

One goal of our skill reconstruction was also to reconstruct the individual differences and similarities between the operators in the style of controlling the crane. For this reason we used traces of two operators that use, at least quantitatively, very different control styles. The two operators are named S and L.

Operator S uses very conservative strategy that is very reliable but requires more time to accomplish the control task. He uses small accelerations in order to avoid large rope oscillations. Operator L, on the other hand, uses faster and more complex strategy. He is

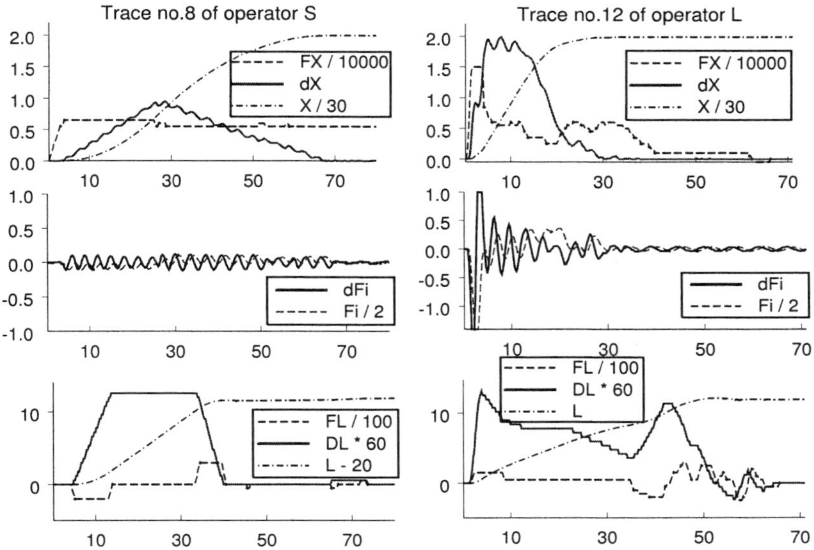


Figure 6.3: Typical traces of operators S and L: on the left is a trace of operator S and on the right a trace of operator L that uses a more complex strategy. Note the differences in the swing of the rope in the second plot from the top.

able to afford large initial swing caused by large acceleration, and later skillfully reduce the swing. Typical traces of both operators are given in Figure 6.3.

We used QUIN to induce qualitative trees from the traces of the two operators. Controlling the crane requires a rule to control the trolley and a rule to control the rope. We induced the desired trolley velocity $\dot{X}_{des}$ and the desired rope length $L_{des}$ in all the following experiments. One reason for this choice of the dependent trajectory variables is to be able to compare the induced qualitative strategies with the strategy induced by GoldHorn in Section 4.2. The choice of the trajectory variables is also discussed in Chapter 7.

Some general observations on the experiments are that all the induced qualitative trees were small and had a low QUIN's error-estimate on the learning set. Since we focus on the interpretability of the induced trees, we manually pruned some of the trees. Besides the presented trees that we were able to interpret, QUIN also induced qualitative trees that are not so straightforward to explain in the light of the crane control task. We give an example of such a tree at the end of Section 6.3.2.

### 6.3.1 Controlling the Rope Length

We used QUIN to induce the desired rope length, i.e. $L_{des}$ from the traces of operator's S and L. $L_{des}$ was the class and the other state variables were the attributes.

From trace no. 8 of operator S a qualitative tree with a single leaf was induced:

$$L_{des} = M^{+}(X) \tag{6.3}$$

Although this is a very simple tree, it is consistent with all the learning examples from the operator's trace. QUIN's consistency of the tree (on the learning set) is 100%. It clearly describes the operator's strategy to control the rope: the operator brings down the load as the trolley moves from its start to the goal position.

From other traces of both operators similar trees were induced. Some of them were more complicated, but all the induced trees had QCF $L_{des} = M^+(X)$ in the root. An example is a qualitative tree induced from another trace of operator S:

$$\begin{aligned} X \le 60.5: &\quad L_{des} = M^+(X) \\ X > 60.5: &\quad L_{des} = M^-(\dot{X}) \end{aligned}$$

In this trace the operator failed to stop the trolley at the exact trolley goal position $X_g$=60. He uses negative trolley velocity to approach $X_g$ from the other direction, but still increases the rope length.

As mentioned before, the operators S and L increase the rope length as the trolley moves from its start to the goal position. However, sometimes the operator reaches the goal position of the trolley first and then (keeping $X$ steady) adjusts the rope length. A qualitative tree with a single QCF $L_{des} = M^+(X)$ is not consistent with such an execution trace. An example is trace no. 12 of the operator L, given in Figure 6.3. QUIN induced from this trace the following qualitative tree:

$$\begin{aligned} \dot{L} \le -0.01: &\quad L_{des} = M^-(\dot{L}) \\ \dot{L} > -0.01: &\quad L_{des} = M^+(X) \end{aligned} \tag{6.4}$$

QCF in the first leaf states that the sign of the rope acceleration is the opposite of the sign of the rope velocity. When the rope velocity is positive (negative), negative (positive) acceleration is used. Therefore QCF $L_{des} = M^-(\dot{L})$ describes the oscillation of the trolley velocity around zero.

When the trolley position $X$ is increasing, the induced tree allows (a) positive rope velocity (by $L_{des} = M^+(X)$) or (b) increasing rope velocity if it is negative (by $L_{des} = M^-(\dot{L})$). Therefore the load is either (a) moving down to approach the goal, or (b) stopping if going in the opposite direction. When the trolley stops, the induced tree allows either (a) zero rope velocity or (b) increasing rope velocity if it is negative. An interpretation of this is that either (a) the load is at the goal position, or (b) approaching the goal from below. Considering the fact that $\dot{L}$=-0.01 is a small negative velocity (one centimeter per second), the possibility (b) can also be interpreted as setting rope velocity to zero when adjusting the load in the goal position.

Comparing the trees of both operators we can conclude that they use the same simple control strategy to control the rope length. They both increase the rope length as the trolley moves towards the goal. Usually this is all it takes to control the rope length. However, both sometimes miss the exact goal position and do some additional adjusting at the end. The trolley position $X$ is the most informative attribute.

### 6.3.2 Controlling the Trolley

Transporting the load from the shore to the ship requires positioning of the trolley above the goal position ($X_g$=60). Therefore the trolley velocity is to be first increased and then decreased, to precisely adjust the trolley at the goal. The task is difficult because it requires that the swing of the rope is zero at the goal. Experienced operators try to reduce the swing by proper accelerations of the trolley.

**Inducing Qualitative Strategy of Operator S**

We used QUIN to induce the desired trolley velocity, i.e $\dot{X}_{des}$ from the traces of the operators S and L. In initial experiments all other state variables were the attributes. However, from many traces, trees that do not correspond to our intuition about controlling the trolley were induced. A simple example is a qualitative tree induced from a trace of operator S:

$$\begin{aligned} L \leq 28.0: &\quad \dot{X}_{des} = M^{+}(L) \\ L > 28.0: &\quad \dot{X}_{des} = M^{-}(X) \end{aligned} \tag{6.5}$$

This tree has a low error-estimate, but implies that the operator adjusts the trolley velocity according to the rope length. The induced qualitative constraints might be true in the execution trace, but are (according to the operator's description of his strategy) not close to human thinking about controlling a crane. In previous experiments (Urbančič, 1994), the operator S never mentioned $L$ as an attribute when describing his skill. However, it is still possible that he might use $L$ as an attribute subconsciously.

There is a simple explanation why QUIN induces such a tree. In the traces $L$ and $X$ behave qualitatively similarly. This is also reflected in the induced $L_{des} = M^{+}(X)$. Therefore, QUIN can express a class's qualitative behaviour using one or the other attribute. From QUIN's point of view, two positively (or negatively) related attributes are equivalent. To induce trees that can be interpreted in the light of human operator's own experience, we decided to leave out the variables $L$ and $\dot{L}$, as discussed also in the next chapter. In all the following experiments the attributes for learning $\dot{X}_{des}$ were $X$, $\Phi$ and $\dot{\Phi}$.

The following experiments with traces of the operator S were a surprise for us. QUIN, in contrary to our expectations, discovered that the operator S is able to reduce the swing of the rope by proper acceleration of the trolley. The operator S controls the crane in a very conservative way and changes the control forces rarely as can be observed in Figure 6.3. His conservative control style with small trolley accelerations results in a small swing of the rope. For this reason, our initial hypothesis was that he does not try to reduce it. Also in experiments with regression trees and GoldHorn, the skill to reduce the swing could not be identified. Rope angle and velocity were typically not important attributes in the induced trees.

One reason that his skill to reduce the swing is hard to identify, is because he uses it rarely, usually just a few times in a trace, sometimes just at the very end. This is what QUIN induced from one of his less successful traces (trace no. 12):

$$\begin{aligned} &X \leq 20.7: \quad \dot{X}_{des} = M^{+}(X) \\ &X > 20.7: \\ &\mid \ X \leq 60.1: \quad \dot{X}_{des} = M^{-}(X) \\ &\mid \ X > 60.1: \quad \dot{X}_{des} = M^{+}(\Phi) \end{aligned} \tag{6.6}$$

This qualitative tree provides a good insight in the operator's control skill. The operator first increases the trolley velocity. At about half distance from the goal ($X$=20.7) he decreases the trolley velocity. At the goal position ($X > 60.1$) he uses the rule $\dot{X}_{des} = M^{+}(\Phi)$. This constraint, telling that the trolley velocity is positively related to the rope angle, describes his skill to reduce the swing of the rope. By the acceleration of the trolley, when the rope angle increases, the angular velocity is decreased. Note that the same qualitative constraint was induced by GoldHorn from a trace of operator L (Section 4.2.1). However, there the constraint applied to the whole execution trace.

The surprising constraint $\dot{X}_{des} = M^+(\Phi)$ lead us to reconsider the instructions that the operator gave through introspection. In (Urbančič and Bratko, 1994) the operator S described his skill of "final balancing" at the goal (when $X > 59.65$) as:

> $F_X$ = 6000. Now $\dot{X}$ oscillates slightly.
>
> When $\dot{X}$ is increasing, a little before the maximum do $F_X$=5500

The operator also gave instructions to accelerate the trolley "a little before the minimum of $\dot{X}$". First note that "$\dot{X}$ oscillates slightly" is the consequence of swinging of the load and that $\dot{X}$ oscillates in the opposite direction than $\dot{\Phi}$. The minima (maxima) of $\dot{X}$ correspond to the points in the middle of the rope swing cycle when $\Phi$ is increasing (decreasing). Considering the fact that the above decreasing of $F_X$ causes negative acceleration, and that "increasing $\dot{X}$ a little before the maximum" corresponds to a point "a little" before the middle of the decreasing $\Phi$ swing cycle, the operators instructions amount to the same effect as the induced QCF: $\dot{X}_{des} = M^+(\Phi)$.

In the same way the operator described his skill to reduce the swing also before reaching the goal, i.e. when $X > 22$ (according to his instructions). This was induced from another one of his traces:

$$\begin{array}{l} X \le 26.9: \quad \dot{X}_{des} = M^+(X) \\ X > 26.9: \\ \mid \quad \Phi \le -0.01: \quad \dot{X}_{des} = M^-(X) \\ \mid \quad \Phi > -0.01: \quad \dot{X}_{des} = M^{-,+,-}(X, \Phi, \dot{\Phi}) \end{array} \tag{6.7}$$

Here the strategy is similar as before, but reducing of the swing starts before reaching the goal ($X > 26.9$). Also, the skill to reduce the swing is here more refined. Let's observe just the part of the tree for $X > 26.9$. One interpretation is that the trolley velocity it to be increased when $\Phi > -0.01$ and $\Phi$ is increasing and $\dot{\Phi}$ is decreasing, i.e. when the rope swings through the vertical to the right. It is interesting that this part of the tree is a refinement of the rule $\dot{X}_{des} = M^+(\Phi)$ and even more precisely corresponds to the operator's instructions to accelerate "a little before the minimum of $\dot{X}$". As explained before, the minimum of $\dot{X}$ corresponds to the point in the middle of the rope swing cycle where $\Phi$ is increasing. QUIN qualified "a little before" as $\Phi > -0.01$.

From the trace no. 8 of the operator S (see Figure 6.3) QUIN induced the following qualitative tree:

$$\begin{array}{l} X \le 20.3: \quad \dot{X}_{des} = M^+(X) \\ X > 20.3: \\ \mid \quad \Phi \le -0.01: \quad \dot{X}_{des} = M^-(X) \\ \mid \quad \Phi > -0.01: \quad \dot{X}_{des} = M^{-,-}(X, \dot{\Phi}) \end{array} \tag{6.8}$$

The second part of the tree $X \le 20.3$ allows increasing of the trolley velocity whenever $\Phi > -0.01$ and $\dot{\Phi}$ is decreasing. This is only partially consistent with the operator's instructions. The increase of the trolley velocity starts "a little before the minimum of $\dot{X}$", i.e. when $\Phi > -0.01$ ($\Phi$ is increasing) but it continues also when $\Phi$ is decreasing as long $\dot{\Phi}$ is decreasing.

This second part, i.e. increasing of the trolley velocity when $\Phi$ is decreasing, is not consistent with other induced rules for reducing the swing of the rope. Our experiments, presented later, show that the rule to reduce the swing is useful, but not very efficient.

### Inducing Qualitative Strategy of Operator L

Besides learning the control strategy of a single operator, we would also like to reconstruct the differences in the individual control styles of the operators S and L. From the trace no. 12 of the operator L the following qualitative tree was induced:

$$\begin{array}{ll} X \leq 29.3: & \dot{X}_{des} = M^{+,+,-}(X, \Phi, \dot{\Phi}) \\ X > 29.3: & \{\text{default rule: } \dot{X}_{des} = M^{-,+}(X, \Phi)\ \} \\ \mid\ \ \dot{\Phi} \leq -0.02: & \dot{X}_{des} = M^{-}(X) \\ \mid\ \ \dot{\Phi} > -0.02: & \dot{X}_{des} = M^{-,+}(X, \Phi) \end{array} \tag{6.9}$$

Similar to the operator S, the operator L first increases the trolley velocity and decreases it later to approach the goal. However, to approach the goal as fast as possible he uses large accelerations that cause a large swing of the rope at the very start. He manages to skillfully reduce this large swing as can be observed in Figure 6.3. In contrast to the operator S, he reduces the swing of the rope also in the first part of the control, while still increasing the trolley velocity. QUIN induced this skill as the QCF $M^{+,+,-}(X, \Phi, \dot{\Phi})$ that applies when $X \leq 29.3$. An interpretation of this QCF is that he decreases the trolley velocity when the rope angle $\Phi$ is decreasing and the rope velocity $\dot{\Phi}$ is increasing. This is, while the rope swings from the vertical to the left, until it reaches it local minimum.

Similarly to the operator S, the operator L also reduces the swing of the rope in the second part of control, that is when the trolley velocity is decreasing. However, he does it more consistently. QUIN induced in the second part of the tree (for $X > 29.3$) the default rule $\dot{X}_{des} = M^{-,+}(X, \Phi)$. This rule, similar to the above mentioned rules, decreases the swing of the rope by increasing the trolley velocity when the rope angle increases. The operator's consistency also enabled QUIN to induce a simpler form (from QUIN's point of view) of this rule. Consider the leaf that applies when $\dot{\Phi} \leq -0.02$. Because $\dot{\Phi}$ is negative, $\Phi$ is decreasing. Since $X$ is increasing (as the trolley moves towards the goal) and $\Phi$ is decreasing, the default rule would require $\dot{X}_{des}$ to decrease. Since $X$ is increasing, a simpler QCF $M^{-}(X)$ achieves the same result as a more complex default rule.

Note that the qualitative tree given by Rule 6.9 is "almost" a generalization of the qualitative strategy (Rule 4.6) obtained by GoldHorn in our previous experiments. Almost, because of the split $\dot{\Phi} \leq -0.02$. If $\dot{\Phi} \leq 0.0$ was induced instead, it would be a generalization of that qualitative strategy.

Finally we also give an example of a qualitative tree we find harder to interpret. This tree was induced from another trace of operator L (trace no. 9):

$$\begin{array}{lll} \dot{\Phi} \leq 0.01 & & \\ \mid \ \Phi \leq 0.11: & \dot{X}_{des} = M^{-}(\dot{\Phi}) & \\ \mid \ \Phi > 0.11: & \dot{X}_{des} = M^{-}(X) & \\ \dot{\Phi} > 0.01: & & \\ \mid \ X \leq 58.6: & \dot{X}_{des} = M^{-,+,-}(X, \Phi, \dot{\Phi}) & \\ \mid \ X > 58.6: & \dot{X}_{des} = M^{+}(\Phi) & \end{array} \tag{6.10}$$

**Comparing Induced Qualitative Strategies of Operators S and L**

By comparing the induced trees we can conclude that both operators use qualitatively similar strategies to control the trolley with some differences. The most important difference is that the operator L reduces the rope swing during the whole trace whereas the operator S tries to reduce the swing only at the end while approaching the goal. One speculation could also be that the operator S is not consistent in reducing the swing: some of the induced trees are consistent with his instructions, others (Rule 6.8) only partially.

### 6.3.3 Transforming Qualitative Trees into Quantitative Strategies

In this section we test how general and successful the induced qualitative trees are by transforming them into operational quantitative strategies. Since controlling the rope length is rather simple, we focus on strategies to control the trolley. We use two qualitative trees of the operator S and one qualitative tree of the operator L, generate qualitatively constrained functions at random, and test the resulting strategies on the simulator. The tree given by Rule 6.6 controls the trolley qualitative equivalently as tested Rule 6.12 (pruned tree of operator S) under the condition that the trolley position does not exceed the goal position ($X \leq 60.1$). This is assured by the task knowledge constraint (the trolley stops at $X_g$) that was used when transforming qualitative trees into quantitative strategies.

To transform the induced qualitative trees into quantitative strategies we use the same procedure as with our experiments with GoldHorn in Section 4.2.3. The reason is to be able to compare the efficiency of the resulting strategies. In Section 4.2.3 we used basic knowledge of the crane control task (task limits, the trolley starts towards the goal, the trolley stops at $X_g$) with a qualitative abstraction of a strategy induced by GoldHorn. The resulting quantitative strategies were expressed using qualitatively constrained functions $f$ and $h$. Functions $f$ map from $[0, 1]$ to $[0, 1]$ whereas functions $h$ map from $[-1, 1]$ to $[-1, 1]$ and cross zero.

By similar derivation as in Section 4.2.3, qualitative tree (Rule 6.9) induced from a trace of the operator L, can be transformed into a set of quantitative strategies:

$$\begin{array}{ll} X \leq 29.3: & \dot{X}_{des} = \dot{X}_{max} f_1^{+}(\frac{X}{29.3}) + k_1 h_1^{+}(\frac{\Phi}{\Phi_{max}}) + k_2 f_2^{-}(\frac{\dot{\Phi} + \dot{\Phi}_{max}}{2\dot{\Phi}_{max}}) \\ X > 29.3: & \\ \mid \ \dot{\Phi} \leq -0.02: & \dot{X}_{des} = \dot{X}_{max} h_2^{-}(\frac{X - X_g}{X_g - 29.3}) \\ \mid \ \dot{\Phi} > -0.02: & \dot{X}_{des} = \dot{X}_{max} h_2^{-}(\frac{X - X_g}{X_g - 29.3}) + k_1 h_3^{+}(\frac{\Phi}{\Phi_{max}}) \end{array}$$

Table 6.1: Efficiency of induced qualitative trees: the table gives the number of successful strategies, average and best time of successful strategies that were obtained from induced qualitative trees. The efficiency of strategies obtained using GoldHorn and strategies that do not consider the rope angle are given for comparison. For each qualitative strategy 90 corresponding quantitative strategies were tested on the simulator.

| Qualitative tree used | Successful strat. (out of 90) | Average time (in seconds) | Best time (in seconds) |
|---|---|---|---|
| Rule 6.8 (operator S) | 26 | 115 | 63 |
| Rule 6.7 (operator S) | 57 | 100 | 53 |
| Rule 6.9 (operator L) | 73 | 84 | 44 |
| GoldHorn strategy Rule 4.6 (operator L) | 75 | 93 | 49 |
| Rule 6.12 (pruned tree) | 8 | 146 | 101 |

Qualitative tree given by Rule 6.8 and induced from a trace of the operator S is transformed into a set of quantitative strategies:

$$\begin{aligned}
&X \leq 20.3: \quad \dot{X}_{des} = \dot{X}_{max} f_1^+(\frac{X}{20.3}) \\
&X > 20.3: \\
&\mid \quad \Phi \leq -0.01: \quad \dot{X}_{des} = \dot{X}_{max} h_1^-(\frac{X - X_g}{X_g - 20.3}) \\
&\mid \quad \Phi > -0.01: \quad \dot{X}_{des} = \dot{X}_{max} h_1^-(\frac{X - X_g}{X_g - 20.3}) + k_2 h_2^-(\frac{\dot{\Phi}}{\dot{\Phi}_{max}})
\end{aligned} \tag{6.11}$$

Qualitative tree (Rule 6.7) induced from another trace of the operator S gives:

$$\begin{aligned}
&X \leq 26.9: \quad \dot{X}_{des} = \dot{X}_{max} f_1^+(\frac{X}{26.9}) \\
&X > 26.9: \\
&\mid \Phi \leq -0.01: \dot{X}_{des} = \dot{X}_{max} h_1^-(\frac{X - X_g}{X_g - 26.9}) \\
&\mid \Phi > -0.01: \dot{X}_{des} = \dot{X}_{max} h_1^-(\frac{X - X_g}{X_g - 26.9}) + k_1 h_2^+(\frac{\Phi}{\Phi_{max}}) + k_2 h_3^-(\frac{\dot{\Phi}}{\dot{\Phi}_{max}})
\end{aligned}$$

These quantitative strategies were tested similarly as strategies $S_{q0}$ in our experiments with GoldHorn in Section 4.2.3: $k_1$ and $k_2$ were set to values $\dot{X}_{max}/2$, $\dot{X}_{max}/4$ and $\dot{X}_{max}/8$. For each $(k_1, k_2)$ combination, 10 sets of piece-wise linear functions $f$ and $h$ were generated at random with $R = 10$, i.e. by using 10 random numbers. We get $3 \times 3 \times 10 = 90$ different quantitative strategies in this way. These strategies were tested on the simulator using Rule 4.5 to control the rope length and system's dynamics learned with LWR from the original trace and another two traces of the same operator.

The results given in Table 6.1 show that all the qualitative trees correspond to successful quantitative strategies. However, the efficiency and the proportion of successful strategies depends on the qualitative tree used. The most successful is qualitative tree (Rule 6.9) that was induced from a trace of the operator L. It is as efficient as the qualitative abstraction of the strategy induced by GoldHorn from the same trace. The strategies (learned from the same trace) are very similar and are both very successful in reducing the swing of the rope. Qualitative trees obtained from traces of the operator S are less efficient. One reason is that he (and the induced strategies) reduces the swing of the rope only in the second part of control when the trolley is slowing down to approach the goal.

Qualitative tree given by Rule 6.8 is the least efficient. Most of the quantitative strategies obtained from this tree approach the goal, but are too slow in reducing the swing of the rope. This led us to question its ability to reduce the swing. For this reason we also tested a pruned version of the tree that does not consider the rope angle:

$$\begin{aligned} X \leq 20.3: &\quad \dot{X}_{des} = M^{+}(X) \\ X > 20.3: &\quad \dot{X}_{des} = M^{-}(X) \end{aligned} \tag{6.12}$$

Out of 90 strategies obtained from this qualitative tree, only 8 were successful, all of them very slow. This is considerably worse than with the unpruned tree that also considers the rope angle. Our explanation is that the tree given by Rule 6.8 can reduce the swing of the rope, but it is not very efficient. Its rule for reducing the swing is only partially correct.

## 6.4 Discussion

We used QUIN in the reconstruction of human control skill of riding a bicycle and controlling a crane. Some interesting points are:

- In both domains QUIN induced qualitative trees that can be explained in the light of the control task and provide good insight in the human control skill. In both domains some surprising or non-trivial aspects of human control skill were induced.

- In the crane domain QUIN was used to induce qualitative control strategies of the two operators that use different control styles. The induced qualitative trees offer a possibility to reconstruct some qualitative similarities and differences in their control styles. One such difference is that one operator decreases the swing of the rope during the whole control while the other operator does it at the end only.

- Qualitative strategies offer a space for controller optimization and were transformed into operational control strategies. The success and efficiency of such controllers also depend on the operator's control style. Since the strategies were induced as (qualitative) constraints on the generalized operator's trajectory, this transformation yields very robust controllers.

- When transforming a qualitative strategy into operational quantitative strategies, we can use some domain knowledge imposing constraints on the tuned parameters. This was also demonstrated in the experiments in the crane and acrobot domains in Chapter 4. The resulting strategies perform the control task in a way that is qualitatively equivalent to the original operator's control style. To achieve the control style that is also quantitatively similar to the operator's, one idea is to use information from the operator's trace. A very simple example is given in the bike domain. We used the maximal and minimal values of the variables from the operator's execution trace to set the definition area of the tuned QCF.

# Chapter 7

# Role of User Knowledge and Intervention

The proposed approach to skill reconstruction provides a framework and a set of tools which make it possible for the user to build a behavioural model taking background and task knowledge and his/her intuitions like causality in the system into account. In this chapter we review the cases where the user knowledge and interventions were used, describe the main principles and give some ideas how user interventions could be automated.

Sections 7.1 and 7.2 describe how the trajectory variables and the induced equations were selected. In Section 7.3 we review the use of domain knowledge used to transform induced qualitative into operational quantitative strategies. Section 7.4 gives the types and the principles of human intervention. At the end we discuss the role of user knowledge and give some ideas how user interventions could be automated.

## 7.1 Selection of Trajectory Variables

A generalized trajectory is a set of constraints among the state variables. We have chosen to represent the induced constraints as a function that expresses a chosen state variable (dependent variable) as a function of other state variables. There are two reasons for adopting such a representation. The first reason is to simplify the computation of the control action, i.e. to minimize the constraint error in a *single* variable. The second reason is comprehensibility. It is up to the user to select dependent and independent variables. When making this selection, we used the following domain knowledge:

- In the experiments in the crane domain (Section 4.2.1 and Section 6.3) $\dot{X}_{des}$ and $L_{des}$ were chosen as the dependent variables. This decision was made according to our notion of the causality in the system, namely, that the force applied to the trolley ($F_X$) most directly affects the trolley velocity $\dot{X}$, and the force applied to the rope ($F_L$) most directly affects the rope velocity $\dot{L}$. However, the desired value of a variable's derivative ($\dot{L}_{des}$ for example) in a particular state can be computed if we have an approximate (learned) model of system's dynamics and the desired value of a variable ($L_{des}$). According to this naive notion of the causality $\dot{L}_{des}$ and $L_{des}$ are both suitable dependent variables. We arbitrarily chose $L_{des}$ between $\dot{L}_{des}$ and $L_{des}$. The decision to use $\dot{X}_{des}$ was similar. In another experiment in this domain (Section 3.3.1) we used $\dot{X}_{des}$ and $\dot{L}_{des}$ as the dependent variables.

  When inducing $\dot{X}_{des}$ we decided to leave out the variables $L$ and $\dot{L}$, since in the previous experiments (Urbančič and Bratko, 1994) operators did not mention variables $L$ and $\dot{L}$

when describing their skill of controlling the trolley. When inducing $L_{des}$ the variable $\dot{L}$ was left out for reasons discussed in Section 7.4.

- In the experiments with the acrobot (Section 4.3.1) $q2_{des}$ was chosen as the dependent variable. According to our notion of the causality in the system, the torque applied to the $q2$ joint most directly affects variable $\dot{q2}$. The decision between $\dot{q2}$ and $q2$ was arbitrary. Choosing $q2_{des}$ enables a comparison of our results with the previous work (DeJong, 1995; Spong, 1994) in this domain. When inducing $q2_{des}$ the variable $\dot{q2}$ was left out because it is a derivative of the dependent variable (discussed in Section 7.4).

- In the experiments with bicycle riding (Section 6.2.1), $\theta_{des}$ was chosen as the dependent variable. According to our notion of the causality in the system, the torque applied to the handlebars most directly affects the angular velocity of the handlebars ($\dot{\theta}$). The decision between $\theta_{des}$ and its derivative was arbitrary.

## 7.2 Selection of Equations

In Chapter 4 we used GoldHorn to induce generalized trajectories from execution traces. Here we describe the use of domain knowledge in selecting among the induced equations.

When GoldHorn induces equations, it sorts them according to their significance judged by normalized deviation and the multiple correlation coefficient. In all the presented experiments with GoldHorn we induced equations with linear and quadratic terms, that is GoldHorn's default value 2 of the parameter *max depth* (depth of constructed terms) that determines the maximal number of variables multiplied in a term. In our initial experiments with GoldHorn we also experimented with equations with terms of higher order, but we did not find them very useful. Often such equations could not be directly used to control the system since the dependent variable could not be expressed as a function of other variables. When they did, the yield and robustness of the resulting controllers was usually not improved. We also found that equations with higher order terms are often more difficult to interpret.

### 7.2.1 Crane: Controlling the Trolley

GoldHorn was used to induce equations using the variables $X$, $\Phi$, $\dot{\Phi}$ and $\dot{X}_{des}$. 20 most significant equations (according to the normalized deviation) are given in Table 7.1. The first equation in the table was ranked first according to both GoldHorn's significance measures (normalized deviation and multiple correlation coefficient). This is the equation that successfully controls the trolley as explained in Section 4.2.1.

In the same way as the most significant equation, we tested the second and the third most significant equations. They are also successful, but much slower. They require 185 and 179 seconds respectively to accomplish the task. The majority of other highly ranked equations cannot be so directly used to control the trolley because they allow two solutions for $\dot{X}_{des}$.

### 7.2.2 Crane: Controlling the Rope

GoldHorn was used to induce equations using the variables $X$, $\dot{X}$, $\Phi$, $\dot{\Phi}$ and $L_{des}$. 20 most significant equations (according to the normalized deviation) are given in Table 7.1. The first

Table 7.1: 20 most significant (according to the normalized deviation) equations that were induced by GoldHorn using variables $X$, $\Phi$, $\dot{\Phi}$ and $\dot{X}_{des}$ (upper table) and variables $X$, $\dot{X}$, $\Phi$, $\dot{\Phi}$ and $L_{des}$ (lower table). According to the multiple correlation coefficient the ranking of the first 10 equations is the same. $\epsilon$ denotes a coefficient with the absolute value less than 0.01.

| Rank | Induced equations for the trolley control |
|---|---|
| 1 | $509.52 + 51.24X + 28.21\Phi - 564.32\dot{X}_{des} - X^2 = 0$ |
| 2 | $445.02 + 52.33X + 4.91\dot{\Phi} - 543.03\dot{X}_{des} - X^2 = 0$ |
| 3 | $444.19 + 52.34X - 542.63\dot{X}_{des} - X^2 = 0$ |
| 4 | $-1.78 + 0.03X + 3.03\dot{X}_{des} - 0.03X\dot{X}_{des} - \dot{X}^2_{des} = 0$ |
| 5 | $-0.05 + 0.35\Phi + 1.83\dot{X}_{des} - 0.02X\Phi - \dot{X}^2_{des} = 0$ |
| 6 | $-0.05 + 1.81\dot{X}_{des} - 0.02X\Phi + 0.31\Phi\dot{X}_{des} - \dot{X}^2_{des} = 0$ |
| 7 | $-923.51 + 74.61X - 320.95\Phi - 79.64\dot{\Phi} - X^2 = 0$ |
| 8 | $-1.73 + 3.11\dot{X}_{des} + \epsilon X^2 - 0.02X\dot{X}_{des} - \dot{X}^2_{des} = 0$ |
| 9 | $-923.66 + 74.61X - 320.86\Phi - X^2 = 0$ |
| 10 | $-0.05 + 1.83\dot{X}_{des} - 0.02X\Phi - 0.1\Phi^2 - \dot{X}^2_{des} = 0$ |
| 11 | $-0.05 + 1.84\dot{X}_{des} - 0.02X\Phi - 0.33\dot{\Phi}^2 - \dot{X}^2_{des} = 0$ |
| 12 | $-0.06 + 2.16\dot{X}_{des} - 0.01X\dot{X}_{des} - 0.43\dot{\Phi}^2 - \dot{X}^2_{des} = 0$ |
| 13 | $-0.06 + 2.09\dot{X}_{des} - 0.01X\dot{X}_{des} - 0.12\Phi^2 - \dot{X}^2_{des} = 0$ |
| 14 | $-0.81 + 2.14\dot{X}_{des} + \epsilon X^2 - 0.02X\Phi - \dot{X}^2_{des} = 0$ |
| 15 | $-0.7 + 0.01X + 2.02\dot{X}_{des} - 0.02X\Phi - \dot{X}^2_{des} = 0$ |
| 16 | $-0.07 + 2.18\dot{X}_{des} - 0.02X\dot{X}_{des} + 0.32\Phi\dot{X}_{des} - \dot{X}^2_{des} = 0$ |
| 17 | $-0.07 + 0.29\Phi + 2.15\dot{X}_{des} - 0.02X\dot{X}_{des} - \dot{X}^2_{des} = 0$ |
| 18 | $-0.07 - 0.39\Phi + 1.8\dot{X}_{des} - 0.22\Phi^2 - \dot{X}^2_{des} = 0$ |
| 19 | $-0.07 + 1.8\dot{X}_{des} - 0.01X\Phi - 0.09\dot{\Phi}\dot{X}_{des} - \dot{X}^2_{des} = 0$ |
| 20 | $-0.07 + 1.86\dot{X}_{des} - 0.01X\Phi - \epsilon X\dot{X}_{des} - \dot{X}^2_{des} = 0$ |

| Rank | Induced equations for the rope control |
|---|---|
| 1 | $-1207.68 + 12.82X + 60.21L_{des} + 0.12X^2 + 0.32X\dot{X} - XL_{des} = 0$ |
| 2 | $-1188.28 + 15.53X - 12.28\dot{X} + 59.50L_{des} + 0.08X^2 - XL_{des} = 0$ |
| 3 | $-1204.19 + 14.73X - 3.57\Phi + 59.52L_{des} + 0.09X^2 - XL_{des} = 0$ |
| 4 | $-1200.52 + 14.34X - 2.40\dot{\Phi} + 59.61L_{des} + 0.10X^2 - XL_{des} = 0$ |
| 5 | $-1202.09 + 14.30X + 59.72L_{des} + 0.10X^2 + 0.06X\Phi - XL_{des} = 0$ |
| 6 | $-1199.75 + 14.37X + 59.58L_{des} + 0.10X^2 + 0.03X\dot{\Phi} - XL_{des} = 0$ |
| 7 | $-1199.85 + 14.37X + 59.58L_{des} + 0.10X^2 - XL_{des} = 0$ |
| 8 | $-0.44 + 0.03X + 20.65\dot{X} + 0.08X\dot{X} - \epsilon XL_{des} - \dot{X}L_{des} = 0$ |
| 9 | $0.21 + 0.01X + 20.68\dot{X} - 0.03L_{des} + 0.08X\dot{X} - \dot{X}L_{des} = 0$ |
| 10 | $1.01 + 20.7\dot{X} - 0.07L_{des} + 0.08X\dot{X} + \epsilon XL_{des} - \dot{X}L_{des} = 0$ |
| 11 | $0.13 + 20.73\dot{X} - 0.03L_{des} + \epsilon X^2 + 0.08X\dot{X} - \dot{X}L_{des} = 0$ |
| 12 | $0.04 + 20.42\dot{X} + 0.33\Phi + 0.08X\dot{X} - 0.21\dot{X}\Phi - \dot{X}L_{des} = 0$ |
| 13 | $0.5 + 20.37\dot{X} + 0.17\Phi - 0.02L_{des} + 0.08X\dot{X} - \dot{X}L_{des} = 0$ |
| 14 | $-0.45 + 20.75\dot{X} + \epsilon X^2 + 0.08X\dot{X} - \epsilon XL_{des} - \dot{X}L_{des} = 0$ |
| 15 | $-0.16 + \epsilon X + 20.53\dot{X} + 0.11\Phi + 0.08X\dot{X} - \dot{X}L_{des} = 0$ |
| 16 | $0.04 + 20.43\dot{X} + 0.14\Phi + 0.08X\dot{X} - \epsilon X\dot{\Phi} - \dot{X}L_{des} = 0$ |
| 17 | $0.04 + 20.53\dot{X} + 0.15\Phi + 0.08X\dot{X} - 0.05\dot{X}\dot{X} - \dot{X}L_{des} = 0$ |
| 18 | $0.04 + 20.42\dot{X} + 0.15\Phi + 0.08X\dot{X} - \epsilon X\Phi - \dot{X}L_{des} = 0$ |
| 19 | $0.04 + 20.43\dot{X} + 0.14\Phi + 0.08X\dot{X} - 0.01\dot{X}\dot{\Phi} - \dot{X}L_{des} = 0$ |
| 20 | $0.04 + 20.43\dot{X} + 0.14\Phi + \epsilon\dot{\Phi} + 0.08X\dot{X} - \dot{X}L_{des} = 0$ |

equation in the table was ranked first according to both GoldHorn's significance measures and gives $L_{des}$ as:

$$L_{des} = (-1207.68 + 12.82X + 0.12X^2 + 0.32X\dot{X})/(X - 60.21) \tag{7.1}$$

This rule can be used to control the rope as explained in Section 4.2.1 and it successfully concludes the task in 54 seconds.

The equations ranked after the most significant equation cannot be so directly used to control the rope. For example, when $L_{des}$ is expressed from the second most significant equation, it gives the function that is not defined for $X = 59.50$.

We can also use some other equation induced by GoldHorn instead of the most significant equation. In Section 4.2.1 we presented a simpler equation to show how the domain knowledge and the analysis of the induced equations can be used. This equation was the most significant among equations of this size, i.e. with at most three terms:

$$-4739.64 + 271.53L_{des} - X^2 = 0 \tag{7.2}$$

This equation is judged considerably worse according to GoldHorn's measures of significance. It does not fit the data as well as the most significant equations. Accordingly, when $L_{des}$ is expressed from Equation 7.2:

$$L_{des} = 0.0037X^2 + 17.46$$

it is clear that it will not accomplish the task, since it violates the crane control task specification. The crane control task requires that at the goal position of the trolley ($X_g = 60$), the rope length should be $L_g = 32$. The equation states that at $X = 60$, $L_{des}$ is 30.7. We can use the task knowledge, i.e. the position of the goal, to correct the induced equation by adding the difference between the induced and the required rope length:

$$L_{des} = 0.0037X^2 + 17.46 + 1.3$$

This equation successfully controls the rope length as explained in Section 4.2.1.

### 7.2.3 Experiments in Acrobot Domain

GoldHorn was used to induce constraints among the variables $q1$, $\dot{q1}$, $qcm$ and $q2_{des}$. The most significant equations (according to the normalized deviation) are given in Table 7.2. The most significant equations fit the data well, but they are just GoldHorn's approximations of an identity that holds in the acrobot system, namely, that the mass center angle $qcm$ is a function of $q1$ and $q2$.

Since any of three variables $q1$, $q2$ or $qcm$ can be approximated well using the other two, we eliminated the equations that used all of the three variables. Useless equations that do not involve $q2_{des}$ were also eliminated. In this way we get the best rated equation (ranked 32 in the Table 7.2) that determines the magnitude of $q2_{des}$:

$$q2_{des} = \pm\sqrt{0.947 + 0.258q1^2 + 0.032q1\dot{q1} - 0.006\dot{q1}^2} \tag{7.3}$$

There are two problems with this equation: first, the sign of the square root is not known, and second, the square root does not have real solutions when its argument is negative. This

Table 7.2: The most significant (according to the normalized deviation) equations that were induced by GoldHorn using variables $qcm$, $q1$, $\dot{q1}$ and $q2_{des}$. According to the multiple correlation coefficient the ranking of the equations is similar.

| Rank | Induced equations in the acrobot domain |
|---|---|
| 1 | $-0.034 + 0.856q1^2 + 0.167q1qcm + 0.283q2qcm - qcm^2 = 0$ |
| 2 | $-0.041 + 1.03q1^2 + 0.328q1q2 + 0.06q2^2 - qcm^2 = 0$ |
| 3 | $-0.016 + 1.006q1^2 + 0.229q1q2 + 0.034q2qcm - qcm^2 = 0$ |
| 4 | $-0.014 + 1.021q1^2 + 0.259q1q2 - 0.017q1qcm - qcm^2 = 0$ |
| 5 | $-0.016 + 1.002q1^2 + 0.248q1q2 - 0.001\dot{q1}qcm - qcm^2 = 0$ |
| 6 | $-0.013 + 0.007q1 + 1.001q1^2 + 0.253q1q2 - qcm^2 = 0$ |
| 7 | $-0.013 - 0.008q2 + 1.002q1^2 + 0.254q1q2 - qcm^2 = 0$ |
| 8 | $-0.014 + 0.006qcm + 1.002q1^2 + 0.254q1q2 - qcm^2 = 0$ |
| 9 | $-0.014 + 1.005q1^2 + 0.256q1q2 - qcm^2 = 0$ |
| 10 | $0.074 + 0.589q1^2 + 0.922q1q2 - 0.651q1qcm - q2qcm = 0$ |
| 11 | $0.012 + 0.937q1^2 - 0.084q2^2 + 0.116q2qcm - qcm^2 = 0$ |
| 12 | $-0.138 - 1.313q1q2 + 1.095q1qcm + 1.681q2qcm - qcm^2 = 0$ |
| ... | ... ... ... |
| 25 | $0.067 + 0.005\dot{q1} - 0.013qcm + 0.862q1qcm - qcm^2 = 0$ |
| 26 | $0.069 + 0.005\dot{q1} + 0.86q1qcm - qcm^2 = 0$ |
| 27 | $0.067 + 0.87q1qcm - 0.004\dot{q1}qcm - qcm^2 = 0$ |
| 28 | $0.069 - 0.008qcm + 0.864q1qcm - qcm^2 = 0$ |
| 29 | $0.07 + 0.863q1qcm - qcm^2 = 0$ |
| 30 | $0.435 - 1.804q1q2 - 0.952q2^2 + 0.073q2qcm - qcm^2 = 0$ |
| 31 | $0.434 - 1.734q1q2 - 0.941q2^2 - qcm^2 = 0$ |
| 32 | $0.947 + 0.258q1^2 + 0.032q1\dot{q1} - 0.006\dot{q1}^2 - q2^2 = 0$ |

happens just inside a small region near the edge of the state space. In order to determine $q2_{des}$ at this edge region and also the sign of $q2_{des}$ elsewhere in the state space, we can consider another equation induced by GoldHorn. We decided to use a simple equation with three terms most. After eliminating useless equations that do not involve $q2_{des}$ and equations where it was not possible to express $q2_{des}$ as a function defined for every possible value of other variables the most significant equation is:

$$0.060 - 0.754q2 - qcm = 0 \tag{7.4}$$

Equations 7.3 and 7.4 are combined into a control rule that successfully concludes the acrobot control task as explained in Section 4.3.1

## 7.3 Using Domain Knowledge with Qualitative Strategies

Qualitative strategies are defined at a higher level of abstraction and thus define a set of quantitative strategies. In our experiments with skill reconstruction using GoldHorn (Chapter 4) and QUIN (Chapter 6) we used the domain knowledge to transform the induced qualitative strategies into operational quantitative strategies as described below.

### 7.3.1 Experiments in Crane Domain

In Section 4.2.2 the qualitative strategy for the trolley control (Rule 4.6) was derived from an equation that GoldHorn induced from an execution trace. In Section 4.2.3 we performed two experiments with the qualitative strategy for the trolley control and experimented with strategies referred to as $S_{q0}$,$S_{q1}$ and $S_{q2}$ to study the effects of using different levels of domain knowledge.

Basic knowledge was used in the first experiment. This consisted of the task limits ($X < X_{max}$, $|\Phi| < \Phi_{max}$, $\left|\dot{X}\right| < \dot{X}_{max}$) and two simple facts of the control task:

- The trolley starts towards the goal: $0 < \dot{X}_{des}(0,0) \leq \dot{X}_{max}$.
- The trolley stops at $X_g$: $\dot{X}_{des}(X_g, 0)$=0.

The stated domain knowledge defines constraints on the qualitative parameters of the induced qualitative strategy. The resulting quantitative strategies are referred to as strategies $S_{q0}$.

More knowledge of the crane control task was used in the second experiment. In addition to the above stated knowledge, the following constraint was also used:

- *Continuous and fast at the point* $X$=$X_{mid}$ ($\dot{X}_{des}(X_{mid}-\epsilon, 0)$=$\dot{X}_{des}(X_{mid}+\epsilon, 0)$ =$\dot{X}_{max}$).

In this experiment the induced QCFs $M^{+,+}(X, \Phi)$ and $M^{-,+}(X, \Phi)$ were restricted to functions with quadratic dependence on $X$ and linear dependence on $\Phi$, i.e. functions $p_{1,2}X^2 + r_{1,2}X + k_{1,2}\Phi + n$ in Rule 4.8. We wanted the resulting quantitative strategies (strategies $S_{q1}$) to be simple. The linear dependence in both variables would be another obvious choice. We used quadratic functions as a tradeoff between simplicity and variety in the resulting strategies. This would not be necessary. All 12 strategies using linear functions only (obtained when $p_{1,2} = 0$) were successful with average time 62 seconds. The stated domain knowledge defines constraints on the qualitative parameters of the induced qualitative strategy. By using an additional constraint requiring that the trolley travels fast we also defined strategies $S_{q2}$.

### 7.3.2 Experiments in Acrobot Domain

In Section 4.3.2 we derived a qualitative strategy for the acrobot swing-up task (Rule 4.15) from two equations that GoldHorn induced from an execution trace. In Section 4.3.3 basic knowledge of the swing-up task was used to transform the qualitative strategy into a set of operational quantitative strategies.

We used knowledge of the control task limits ($|q1| \leq \pi$, $\left|\dot{q1}\right| \leq 20$, $|q2| \leq q2_{max}$) and the requirement that $q2_{des}$ is a continuous function. We also used some knowledge about "reasonable" values of the qualitative parameters. However, these "reasonable" values of qualitative parameters can be determined from the quantitative strategy induced by GoldHorn. For example, qualitative term *smallQcm* was concretised into values 0.0, 0.05 or 0.1, which are close to the its value from the induced quantitative strategy.

### 7.3.3 Experiments with Bicycle Riding

In Section 6.2.1 we used QUIN to induce a qualitative strategy for the bicycle riding task (Rule 6.1). The transformation of the qualitative strategy into quantitative strategies was done using real-valued functions $f$ that respect the qualitative constraints of QCF $\theta_{des} = M^{+,+,-}(\omega, \dot{\omega}, \Psi_g)$. To set the definition area of $f$, the maximal and minimal values from the operator's trace were used. $\theta_{min}$ and $\theta_{max}$ that define the range of $f$ were randomly selected from the minimal and maximal values of $\theta$ in different traces. This random selection is not needed if we use $\theta_{min}$ and $\theta_{max}$ from the original operator's trace. We used some domain knowledge to define additional constraints on $f$:

- *drive straight if the bike is aiming at the goal*: $f(0,0,0) = 0$.
- *balancing is more important than aiming at the goal*: when $\omega$ and $\dot{\omega}$ both change from their minimal to their maximal values, $\theta_{des}$ changes $c$ times ($c > 1$) of the $\theta_{des}$ change when $\Psi_g$ changes from its minimal to its maximal value. Each of the tested values $c = 2, 3, 4, 5$ yielded successful controllers.

## 7.4 Types of Human Intervention

The types and principles of user knowledge and interventions used in the experiments are:

- The dependent trajectory variables were selected manually according to the domain knowledge, i.e. by selecting a state variable that appeared to be close to the human intuition about the system's control considering a naive notion of the causality in the system. Among the state variables we choose a variable that, according to the human notion of causality, appears to be the most directly affected by the control action. By choosing a dependent variable, we make two assumptions. First, we assume that the system can be controlled directly through the dependent variable, i.e. that it is possible to do the desired control task by applying the appropriate changes to the dependent variable. Second, we assume that it is possible to (sufficiently well) determine the desired value of the dependent variable as a function of other state variables excluding the derivatives of the dependent variable. We exclude the derivatives of the dependent variable according to our notion of causality: if controlling a variable requires considering its derivative, then it is more intuitive to control the derivative (and *it* should be the dependent variable) since the derivative affects the variable and not vice versa.
- When inducing the constraints defining the generalized trajectory, we can leave out some state variables to keep the constraints as simple as possible. In the case of crane rope control we left out the variables that the operators did not mention when describing their skill. In the experiments with GoldHorn we left out the derivative of the dependent trajectory variable for the reason stated in the previous paragraph.
- When selecting equations induced by GoldHorn we disregarded useless equations that reflect trivial identities in the system. Sometimes a single equation cannot be directly used to control the system since it does not uniquely define the value of the dependent variable as discussed also in Section 7.5.2. In such a case, additional equations can be used to determine the value of the dependent variable. We presented an example of combining the induced equations in the acrobot domain.
- In the crane domain we used the task knowledge to correct coefficients in the induced equation for the rope control. A simple goal analysis was used to determine the appropriate value of a coefficient in the equation.

- Domain knowledge was used in transforming the qualitative into quantitative strategies. The motivation for its use is to increase the chances of success of a random controller that respects the induced and domain constraints and gain controllers with the desired properties. As shown in the crane domain, different levels of domain knowledge can be used for this purpose. The types of domain knowledge used are: a) maximal and minimal values of variables, b) the requirement that the generalized trajectory be continuous, c) task knowledge such as the requirement to start in the direction of the goal, stop at the goal, or drive straight if going in the correct direction, and d) additional domain knowledge such as: travel fast, balancing is more important than goal aiming and "reasonable" values of qualitative parameters. To test different quantitative strategies that respect the induced qualitative and the domain knowledge constraints, we decided to use (randomly generated) piece-wise linear functions for their simplicity.

## 7.5 Discussion

The proposed problem decomposition, GoldHorn, QUIN and the transformation of qualitative into quantitative strategies involve possible choices among alternatives. The user decides which variables might be "close to human thinking", makes choices among alternatives offered by GoldHorn and QUIN, and decides which domain knowledge to use with induced qualitative strategies to get controllers with the desired properties.

The approach provides a framework and a set of tools which make it possible for the user to take background and task knowledge, and his/her notion of the causality in the system into account. Some of this knowledge, such as choosing the dependent trajectory variable, is required and some optional. The previous section describes the types of the user knowledge and interventions that can be used within this framework. It is up to the user to make these choices and intelligently experiment with them. In the following paragraphs we comment on the required user competence, and give some ideas how these choices could be automated.

### 7.5.1 Comment on User Competence Required

All the experiments in this thesis were carried out by the author, including all the "user interventions". This invites the question: Was the success of the user interventions not just due to the extraordinary competence of "the user" in this case? What would happen if a less expert made these interventions? Does the success critically depend on user competence?

There is some empirical evidence that it does not. Several students experimented with the original approach to behavioural cloning and with the decomposition approach of this thesis. Those using the decomposition approach invariably achieved a much better success. Siegl (1995; crane domain), Juvan (2000; crane domain), Miletić (1997; acrobot domain), Blažina et al. (1996; acrobot domain) and Vladušič (2000; acrobot domain) used the original approach. They consistently report on very low yield (below 5%) of successful controllers. Bratuša (1998; crane domain) and Javornik (1998; pole-cart domain) used the decomposition approach (with GoldHorn) and were consistently getting incomparably higher yield (over 70%).

### 7.5.2 Using Equations

To use a single equation to control a system by minimizing the error in the dependent variable the equation has to satisfy two conditions. First, the equation must express the dependent variable as a function of other variables from the equation. For example, suppose $Y_{des}$ is the dependent variable. The equation $k_0 + k_1 X + {Y_{des}}^2 = 0$ does not satisfy this condition since it does not define the sign of $Y_{des}$. Second, the dependent variable, as a function of other variables, must be defined for every possible combination of the values of its arguments (independent variables). For example, suppose $Y_{des}$ is the dependent variable and $\Phi$ can take the value 0. The equation $k_0 + k_1 X + k_2 \Phi - \Phi Y_{des} = 0$ does not satisfy this condition since the corresponding $Y_{des}$ function is not defined for $\Phi = 0$.

The constraint to induce only such *well defined* dependent variable functions cannot be given to GoldHorn. One alternative would be to use the equation discovery system Lagramge (Todorovski and Džeroski, 1997; Todorovski, 1998) instead of GoldHorn. Lagramge enables the use of context free grammars to constraint the search space of possible equations. In this way we could reduce the search space and possibly induce only well defined functions of the dependent variable.

Another alternative is to use a *combination* of several equations to define a well defined dependent variable function since all of the induced equations reflect the constraints between the state variables in the execution trace. The choice which equations to use could be made according to some significance or simplicity measures.

An interesting idea would also be to automate equations analysis presented on the example of the crane rope control. Such a system would consider the given domain knowledge to eliminate or even correct equations.

### 7.5.3 Choice of Trajectory Variables

One reason to represent the induced generalized trajectory as a function expressing a chosen state variable (dependent variable) as a function of other state variables is to simplify the computation of the control action, i.e. to minimize the constraint error in a *single* variable. We used the human notion of causality in the system to select the dependent variable. A similar use of the domain knowledge is also in CHURPs (Stirling, 1995), where the user-supplied influence matrix describes the relationships between the controls and the effects of the controls.

An alternative that does not require to select the dependent variable is to minimize the constraint error in all the state variables as explained in Section 3.1.1. For example, when inducing equations GoldHorn does not require the specification of the dependent variable. The system could be controlled according to the induced equation by selecting an action that minimizes the equation error in all the variables according to Equation 3.2. However, this requires the definition of the constraint error in all the state variables and opens the question whether to weight the errors in different variables or treat the variables as equal.

In the crane and acrobot experiments we left out variables that can be approximated well by using other variables. This might sometimes reduce the complexity of the generalized trajectory. The decision to leave out such a variable depends of course on the learning method used. When using GoldHorn, using a such variable often reduces the learning error of the generalized trajectory, but it also increases its complexity and the generalization error. On the other hand, using such a variable with decision trees might reduce the complexity of the induced tree. The decision which variables to leave out was done manually by considering the domain knowledge, but it might also be inferred as explained in the following example from the crane domain.

When learning $\dot{X}_{des}$ trajectory, we decided to leave out variables $L$ and $\dot{L}$ and therefore used some extra knowledge. However, the most significant equation that GoldHorn induces using all the variables ($X$, $\dot{X}_{des}$, $\Phi$, $\dot{\Phi}$, $L$, $\dot{L}$) does not involve $\dot{X}_{des}$ at all:

$$-1197.42 + 14.34X + 59.62L + 0.10X^2 - XL = 0$$

This equation fits the data nearly perfectly and shows that $L$ can be approximated well as a function of $X$. Therefore $L$ should be left out according to the argument above. One could use the following reasoning to leave out also $\dot{L}$. Since (a) $L$ can be approximated well as a function of $X$ and (b) $L$ at the next time step can be approximated well as a function of $X$ and $\dot{X}$ (as induced in Equation 7.1) then $\dot{L}$ can also be approximated well as the difference of these two functions and it should be left out.

*Machine Reconstruction of Human Control Strategies*
*D. Šuc*
*IOS Press, 2003*


# Chapter 8

# Conclusions and Further Work

## 8.1 Conclusions

This thesis addresses problems that were generally observed with the original approach to skill reconstruction by machine learning.

We proposed a decomposition of the learning problem into two learning problems: the induction of a generalized trajectory, and the learning of dynamics. This gives rise to goal-directed strategies that consider system's dynamics. This seems to improve the robustness of resulting controllers with respect to the learning error and with respect to changes in the control task. The proposed problem decomposition is general and not tied to a particular learning technique. It also provides a framework that enables the use of background and task knowledge.

The transparency of the induced strategies, as models of human control skill, is improved in two ways. First, by the induction of *symbolic constraints* that define generalized operator's trajectory. The generalized trajectory can be seen as a continuously changing subgoal. Second, by learning of *qualitative* strategies. We demonstrated that qualitative strategies can capture important and non-trivial aspects of human control skill.

Experiments with qualitative strategies opened new perspectives to the reconstruction of human control skill. These include the possibility of qualitative reasoning about human control skill, reconstruction of individual differences in operators's control styles and the possibility to optimize the operator's skill.

The contributions of the thesis are:

- A decomposition of the behavioural cloning problem into two learning problems: the learning of constraints on the operator's control trajectories and the learning of the system's dynamics. In the experiments carried out in the thesis, the decomposition improves the yield of cloning and the robustness of induced clones. We give a plausible explanation of why this decomposition works better than the original approach where a controller is induced as a direct mapping from states to actions. The framework facilitates the domain knowledge.

- Modelling the operator's control skill with symbolic, and especially qualitative constraints that are induced from operator's execution traces. The induced constraints define the generalized operator's trajectory.

- Experimental evaluation of the proposed decomposition in several dynamic domains. We show that the induced generalized operator's trajectories provide good insight into the operator's control skill and give rise to goal-directed and robust controllers.
- Development of an algorithm for learning of qualitative constraint trees from numerical examples. The algorithm partitions the attribute space to find qualitative functional constraints for each subspace and can handle noisy data.
- Applying the developed algorithm to induce qualitative models of human control skill. Such qualitative strategies define spaces for controller optimization and can be transformed into operational controllers. We give a method for combining domain knowledge of the control task with the induced qualitative strategy and show that using simple facts about the control task improves the success of the resulting controllers.

## 8.2 Further Work

We propose the following directions for further research and improvements of the proposed methods:

- The QUIN algorithm was successfully used in skill reconstruction. We believe, it can be used with other tasks where qualitative models are preferred. The algorithm could be improved in many ways, for example, to use a bottom-up approach instead of a greedy top-down approach. One idea is to first find the subspaces in the attribute space that are consistent with qualitatively constrained functions and then join the overlapping subspaces according to some error measure.
- Qualitative control strategies provide an insight into the operator's control skill. To automate the explanation of the induced strategies, the qualitative simulation could be used. Such an interpreter of qualitative strategies could take the causality or a qualitative model of the system's dynamics into account to generate a set of possible qualitative behaviours of the controlled system.
- Qualitative control strategies provide a space for controller optimization. To reduce the space of admissible controllers the domain knowledge or the qualitative reasoning can be used. An interesting possibility would be to automate qualitative or semi-qualitative reasoning that can verify whether, or under which conditions a qualitative strategy might be successful.
- In our experiments with inducing the operator's generalized trajectories we manually selected the dependent trajectory variable. One reason to select the dependent variable is to simplify the computation of the control action, i.e. to minimize the constraint error in a *single* variable. An alternative that does not require the selection of the dependent variable is to minimize the constraint error in all the state variables. However, this requires a definition of the constraint error in all the state variables which seems to be difficult without the background knowledge.

# References

Aha, D., Kibler, D., and Albert, M. (1991). Instance-based learning algorithms. *Machine Learning*, 6:37–66.

Atkeson, C., Moore, A., and Schaal, S. (1997a). Locally weighted learning. *Artificial Intelligence Review*, 11:11–73.

Atkeson, C., Moore, A., and Schaal, S. (1997b). Locally weighted learning for control. *Artificial Intelligence Review*, 11:75–113.

Bain, M. and Sammut, C. (1999). A framework for behavioural cloning. In Furukawa, K., Michie, D., and S., M., editors, *Machine Intelligence 15*, pages 103–129. Oxford University Press.

Barto, A., Sutton, R., and Anderson, C. (1983). Neuronlike adaptive elements that can solve difficult learning control problems. *IEEE Transactions on Systems, Man and Cybernetics*, SMC-13(5):834–846.

Bertsekas, D. (1987). *Dynamic Programming: Deterministic and Stochastic Models*. Prentice Hall, Englewood Cliffs.

Blažina, I., Miletić, E., and Raguž, I. (1996). Experiments in acrobot control. Students' project report, Faculty of Electrical Engineering and Computer Science, University of Ljubljana, Slovenia.

Boone, G. (1997). Efficient reinforcement learning: Model-based acrobot control. In *International Conference on Robotics and Automation, Albuquerque, NM*.

Bortoff, S. (1992). Pseudolinearization of the acrobot using spline functions. Ph.D.T., University of Illinois at Urbana-Champaign, Dept of Electrical and Computer Engineering.

Bratko, I. (1995). Deriving qualitative control for dynamic systems. In Furukava, K., Michie, D., and Muggleton, S., editors, *Machine Intelligence 14*, pages 367–386. Oxford University Press.

Bratko, I., Mozetič, I., and Lavrač, N. (1989). *KARDIO: a Study in Deep and Qualitative Knowledge for Expert Systems*. MIT Press.

Bratko, I., Muggleton, S., and Varšek, A. (1991). Learning qualitative models of dynamic systems. In *Proceedings of the 8th International Workshop on Machine Learning*.

Bratko, I. and Urbančič, T. (1999). Control skill, machine learning and hand-crafting in controller design. In Furukawa, K., Michie, D., and S., M., editors, *Machine Intelligence 15*, pages 131–153. Oxford University Press.

Bratko, I., Urbančič, T., and Sammut, C. (1995). Behavioural cloning: phenomena, results and problems. In *Automated Systems Based on Human Skill, IFAC Symposioum*. Berlin.

Bratko, I., Urbančič, T., and Sammut, C. (1998). Behavioural cloning of control skill. In Michalski, R., Bratko, I., and Kubat, M., editors, *Machine Learning and Data Mining: Methods and Applications*, pages 335–351. John Wiley & Sons, Ltd.

Bratuša, R. (1998). Reconstruction of crane control skill. B.Sc. Thesis, Faculty of Computer and Information Science, University of Ljubljana, Slovenia.

Brodie, M. and DeJong, G. (1999). Learning to ride a bicycle using iterated phantom induction. In Bratko, I. and Džeroski, S., editors, *Proceedings of the 16th International Conference on Machine Learning*, pages 57–66. Morgan Kaufmann.

Camacho, R. (1994). Laboratory note: On robustness test on induced auto-pilots from single flight plan missions. Available from: http://www.ncc.up.pt/ tau.

Camacho, R. (1995). Using machine learning to extract models of human control skill. In *Proceedings of AIT'95*. Brno, Czech Republic.

Camacho, R. (1998). Inducing models of human control skills. In *Proceedings of the 10th European Conference on Machine Learning*. Chemnitz, Germany.

Camacho, R. (2000). *Inducing Models of Human Control Skills using Machine Learning Algorithms*. PhD thesis, Engineering Faculty of Porto University, Portugal.

Chambers, R. and Michie, D. (1969). Man-machine co-operation in a learning task. *Computer Graphics: Techniques and Applications*, pages 179–186.

Clark, P. and Matwin, S. (1993). Using qualitative models to guide inductive learning. In *Proceedings of the 10th International Conference on Machine Learning*, pages 49–56.

Cleveland, W. (1979). Robust locally-weighted regression and smoothing scatterplots. *Journal of the American Statistical Association*, 74:829–836.

Coiera, E. (1989). Generating qualitative models from example behaviours. Technical Report Technical report 8901, University of New South Wales.

DeJong, G. (1994). Learning to plan in continuous domains. *Artificial Intelligence*, 65:71–141.

DeJong, G. (1995). A case study of explanation-based control. In *Proceedings of the 12th International Conference on Machine Learning*.

Donaldson, P. (1964). Error decorrelation studies on a human operator performing a balancing task. *Med. Electron. Biol. Engng.*, 2:393–410.

Doyle, R. (1988). *Hypothesizing Device Mechanisms: Opening Up the Black Box*. PhD thesis, Massachusetts Institute of Technology.

Duda, R. and Hart, P. (1973). *Pattern classification and scene analysis*. John Wiley & Sons, New York.

Džeroski, S. and Todorovski, L. (1993). Discovering dynamics. In *Proceedings of the 10th International Conference on Machine Learning*, pages 97–103. Morgan Kaufmann.

Falkenhainer, B. and Forbus, K. (1990). Self explananatory simulations: an integration of qualitative and quantitative knowledge. In *Proceedings of the 4th International Workshop on Qualitative Physics*. Lugano, July 1990.

Falkenhainer, B. and Michalski, R. (1986). Integrating quantitative and qualitative discovery: The abacus system. *Machine Learning*, pages 403–422.

Forbus, K. (1984). Qualitative process theory. *Artificial Intelligence*, 24:85–168.

Franklin, G., Powell, J., and Emami-Naeini, A. (1986). *Feedback Control of Dynamic Systems*. Addison Wesley, Reading, MA.

Grobelnik, M. (1994). Container crane simulator.

Hunt, K., Sbarbaro, D., Źbikovski, R., and Gawthrop, P. (1992). Neural netowrks for control systems: A survey. *Automatica*, 28(6):1083–1112.

Javornik, M. (1998). Skill modelling through reconstruction of operator's trajectories. B.Sc. Thesis, Faculty of Computer and Information Science, University of Ljubljana, Slovenia.

John, H. G., Kohavi, R., and K., P. (1994). Irrelevant features and the subset selection problem. In Cohen, W. and Hirsh, H., editors, *Proceedings of the 11th International Conference on Machine Learning*, pages 121–129. Morgan Kaufmann.

Juvan, P. (2000). Behavioural cloning of crane control skill by means of fuzzy regression trees. B.Sc. Thesis, Faculty of Computer and Information Science, University of Ljubljana, Slovenia.

Kaelbling, L., Littman, M., and Moore, A. (1996). Reinforcement learning: A survey. *Journal of Artificial Intelligence Research*, 4:237–295.

Kerr, R. and Kibira, D. (1994). Intelligent reactive scheduling by human learning and machine induction. In *IFAC Symposium On Intelligent Manufacturing*. Vienna.

Kira, K. and Rendell, L. (1992). A practical approach to feature selection. In Sleeman, D. and Edwards, P., editors, *Proceedings of the International Conference on Machine Learning*, pages 249–256. Morgan Kaufmann.

Koehn, B. and Zytkow, J. (1986). Experimenting and theorizing in theory formation. In *Proc. of the International Symposium on Methodologies for Intelligent Systems*, pages 296–307. ACM SIGART Press.

Kononenko, I. (1994). Estimating attributes: Analysis and extensions of relief. In De Raedt, L. and Bergadano, F., editors, *Proceedings of the European Conference on Machine Learning*, pages 171–182. Springer Verlag.

Križman, V. (1998). *Avtomatsko odkrivanje struktore modelov dinamičnih sistemov*. PhD thesis, Faculty of Computer and Information Sciences, University of Ljubljana. in Slovene.

Križman, V., Džeroski, S., and Kompare, B. (1995). Discovering dynamics from measured data. *Electrotechnical Review*, 62:191–198.

Kuipers, B. (1986). Qualitative simulation. *Artificial Intelligence*, 29:289–338.

Kuipers, B. and Åström, K. (1994). The composition and validation of heterogeneous control laws. *Automatica*, 30:233–249.

Langley, P., Simon, H., Bradshaw, G., and Zytkow, J. (1987). *Scietific Discovery: Computational Explorations of the Creative Process*. MIT Press, Cambridge, Massachusetts.

Lee, C. (1990). Fuzzy logic in control systems: Fuzzy logic controller - part 1, part 2. *IEEE Transactions on Systems, Man and Cybernetics*, 20(2):404–435.

Leech, W. (1986). A rule based process control method with feedback. In *Proceedings of the ISA/86 International Conference and Exhibit*. Research Triangle Park, NC 27709: The Instrumentation Society of America.

Makarovič, A. (1991). A qualitative way of solving the pole balancing problem. *Machine Intelligence 12*, pages 241–258.

Michie, D. (1986). Machine learning and knowledge acquisition. In Michie, D. and Bratko, I., editors, *Expert Systems: Automating Knowlege Acquisition*, pages 7–19. Addison-Wesley.

Michie, D. (1993). Knowledge, learning and machine intelligence. In Sterling, L., editor, *Intelligent Systems*, pages 2–19, New York. Plenum Press.

Michie, D. (1995). Problem decomposition and the learning of skills. In Wrobel, S. and Lavrač, N., editors, *Proceedings of the European Conference on Machine Learning*, pages 17–31. Springer.

Michie, D., Bain., M., and Hayes-Michie, J. (1990). Cognitive models from subcognitive skills. In Grimble, M., McGhee, J., and Mowforth, P., editors, *Knowledge-Based Systems in Industrial Control*, pages 71–99. Peter Peregrinus.

Michie, D. and Camacho, R. (1994). Building symbolic representations of intuitive real-time skills from performance data. In Furukawa, K., Michie, D., and S., M., editors, *Machine Intelligence 13*, pages 385–418. Oxford University Press.

Michie, D. and Chambers, R. (1968). Boxes: an experiment in adaptive control. In Dale, E. and Michie, D., editors, *Machine Intelligence 2*, pages 137–152. Edinburgh University Press.

Michie, D. and Sammut, C. (1995). Behavioural clones and cognitive skill models. In Furukava, K., Michie, D., and Muggleton, S., editors, *Machine Intelligence 14*, pages 387–395. Oxford University Press.

Miletić, E. (1997). Behavioural cloning of acrobot control skill. B.Sc. Thesis, Faculty of Electrical Engineering and Computer Science, University of Ljubljana, Slovenia.

Mitchell, T. (1997). *Machine Learning*. The McGraw-Hill Companies, Inc., New York.

Muggleton, S. (1992). *Inductive Logic Programming - Logic Based Approach to ML*. Academic Press, San Diego, Ca. the A.P.I.C. series, no.38.

Quinlan, J. and Rivest, R. (1989). Inferring decision trees using the minimum description length principle. *Information and Computation*, 80:227–248.

Randløv, J. and Alstrøm, P. (1998). Learning to drive a bicycle using reinforcement learning and shaping. In *Proceedings of the 15th International Conference on Machine Learning*, pages 463–471. Morgan Kaufmann.

Renders, J. and Nordvik, J. (1992). Genetic algorithms for process control: A survey. In *Preprints of the IFAC/IFIP/IMACS International Symposium on Artificial Intelligence in Real-Time Control*, pages 579–584.

Richards, B., Kraan, I., and Kuipers, B. (1992). Automatic abduction of qualitative models. In *Proceedings of the National Conference on Artificial Intelligence*. AAAI/MIT Press.

Rissanen, J. (1978). Modelling by shortest data description. *Automatica*, 14:465–471.

Sammut, C. (1994). Recent progress with BOXES. In Furukava, K., Michie, D., and Muggleton, S., editors, *Machine Intelligence 13*, pages 363–383. Oxford University Press.

Sammut, C. (1996). Automatic construction of reactive control systems using symbolic machine learning. *Knovledge Engineering Review*, 11(1):27–42.

Sammut, C., Hurst, S., Kedzier, D., and Michie, D. (1992). Learning to fly. In *Proceedings of the 9th International Workshop on Machine Learning*, pages 385–393. Morgan Kaufmann.

Schaal, S. and Atkenson, C. (1994). Robot juggling: An implementation of memory-based learning. *Control Systems Magazine*, 14:57–71.

Siegl, M. (1995). Behavioural cloning of crane control skill. B.Sc. Thesis, Faculty of Electrical Engineering and Computer Science, University of Ljubljana, Slovenia.

Spong, M. (1994). Swing up control of the acrobot. In *IEEE International Conference on Robotics and Automation*. Morgan Kaufmann. San Diego.

Stanfill, C. and Waltz, D. (1986). Toward memory-based reasoning. *Communications of the ACM*, 29(12):1213–1228.

Stirling, D. (1995). *CHURPs: Compressed Heuristic Universal Reaction Planers*. PhD thesis, Basser Department of Computer Science, University of Sydney.

Sutton, R. (1990). Integrated arhitectures for learning, planning and reacting based on approximating dynamic programming. In *Proceedings of the 7th International Conference on Machine Learning*, pages 216–224. Morgan Kaufmann.

Sutton, R. (1996). Generalization in reinforcement learing: Successful examples using sparse coarse coding. In *Neural Information Processing Systems 8*. MIT Press, Reading, MA.

Sutton, R. and Barto, A. (1998). *Reinforcement Learning: An Introduction*. Adaptive Computation and Machine Learning. MIT Press.

Todorovski, L. (1998). Declerative bias in equation discovery. M.Sc. Thesis, Faculty of Computer and Information Science, University of Ljubljana, Slovenia.

Todorovski, L. and Džeroski, S. (1997). Declerative bias in equation discovery. In *Proceedings of the 14th International Conference on Machine Learning*, pages 376–384, San Mateo, CA. Morgan Kaufmann.

Urbančič, T. (1994). *Avtomatizirana sinteza znanja za vodenje sistemov*. PhD thesis, Faculty of Computer and Information Sciences, University of Ljubljana.

Urbančič, T. and Bratko, I. (1994). Reconstructing human skill with machine learning. In Cohn, A., editor, *Proceedings of the 11th European Conference on Artificial Intelligence*, pages 498–502. John Wiley & Sons, Ltd.

Varšek, A., Urbančič, T., and Filipič, B. (1993). Genetic algorithms in controller design and tuning. *IEEE Transactions on Systems, Man and Cybernetics*, 23(6):1330–1339.

Vladušič, D. (2000). Behavioural cloning of acrobot by means of fuzzy regression trees. B.Sc. Thesis, Faculty of Computer and Information Science, University of Ljubljana, Slovenia.

Šuc, D. (1998). Skill modelling through symbolic reconstruction of operator's trajectories. Master Th., Faculty of Computer and Information Sc., University of Ljubljana, Slovenia.

Šuc, D. and Bratko, I. (1997a). Skill modelling through symbolic reconstruction of operator's trajectories. In Brandt, D., editor, *Automated systems based on human skill : preprints of the 6th IFAC Symposium*, pages 35–38. Kranjska gora, Slovenia.

Šuc, D. and Bratko, I. (1997b). Skill reconstruction as induction of LQ controllers with subgoals. In Mellish, C. S., editor, *Proceedings of the 15th International Joint Conference on Artificial Intelligence*, volume 2, pages 914–920. Nagoya, Japan.

Šuc, D. and Bratko, I. (1999a). Modelling of control skill by qualitative constraints. In *Proceedings of the 13th International Workshop on Qualitative Reasoning*, pages 212–220. University of Aberystwyth. Loch Awe, Scotland.

Šuc, D. and Bratko, I. (1999b). Symbolic and qualitative reconstruction of control skill. *Electronic Transactions on Artificial Intelligence, Section B*, 3:1–22. http://www.ep.liu.se/ej/etai/1999/002/.

Šuc, D. and Bratko, I. (2000). Skill modelling through symbolic reconstruction of operator's trajectories. *IEEE Transaction on Systems, Man and Cybernetics, Part A*, 30(06):617–624.

Williams, B. (1990). Interaction-based invention: designing devices from first principles. In *Proceedings of the 4th International Workshop on Qualitative Physics*. Lugano, July 1990.